PLUMBING IN THE HOUSE

Ernest Hall started work as a junior clerk/trainee sanitary
inspector, and in the course of his career in local government
has worked as housing manager, public health inspector and
public relations officer. He has written numerous articles on
d-i-y and contributed to *D-I-Y Magazine*, *Practical Householder* and *Homeowner*. His publications include *Home
Plumbing* (Newnes Butterworth), *Beginner's Guide to
Domestic Plumbing* (Newnes Butterworth), and *Plumbing*
(Pelham Books Ltd). He has also contributed the plumbing
sections to the Reader's Digest *Book of Home Improvements*
and *Concise Repair Manual*, St Michael's *D-I-Y Book* and
Orbis' *Know How Part Work Manual*.

TEACH YOURSELF BOOKS

PLUMBING
IN THE HOUSE

Ernest Hall
F.R.S.H.

TEACH YOURSELF BOOKS
Hodder and Stoughton Ltd

First Printed 1981

Copyright © 1981
Ernest Hall

Published in the USA by David McKay & Co. Inc., 750 Third Avenue,
New York, NY 10017, USA.

ISBN 0 340 25959 0

Printed and bound in Great Britain for Hodder and Stoughton
paperbacks, a division of Hodder and Stoughton Ltd, Mill Road,
Dunton Green, Sevenoaks, Kent, (Editorial Office; 47 Bedford
Square, London, WC1B 3DP) by
Richard Clay (The Chaucer Press) Ltd, Bungay, Suffolk

Contents

Illustrations		vii
Introduction		xi
1	The Plumber's Tools	1
2	Domestic Cold Water Supply	4
3	Domestic Hot Water Supply	17
4	Fittings Used in Hot and Cold Water Supply	45
5	The Lavatory Suite	65
6	Hard, Soft and Corrosive Water Supplies	77
7	Frost and the Plumbing System	90
8	Above-ground Drainage	104
9	Underground Drainage	125
10	Rural Water Supply and Drainage Problems	140
11	Some Plumbing Techniques	158
12	Plumbing Work in the Kitchen	189
13	Plumbing Work in the Bathroom	211
Glossary		237
Index		251

List of Illustrations

Fig. 1   Water authority stop-cock and service pipe   5
Fig. 2   Householder's main stop-cock and drain-cock   7
Fig. 3   Direct cold water system   8
Fig. 4   Indirect cold water system   9
Fig. 5   The connections to a cold water storage cistern   14
Fig. 6   'Cylinder' storage hot water systems   18
Fig. 7   *Micromet* crystals suspended in cold water storage cistern   21
Fig. 8   An indirect hot water system   22
Fig. 9   An indirect packaged plumbing system   25
Fig. 10   Secondary circulation of hot water   26
Fig. 11   Towel rail circulation connected to direct hot water system   27
Fig. 12   Alternative arrangement of towel rail circulation   28
Fig. 13   Reversed circulation – cause and cure   29
Fig. 14   Single pipe circulation – cause and cure   30
Fig. 15   An off-peak electric water heater   33
Fig. 16   A solar heating system   35
Fig. 17   An instantaneous gas water heater   40
Fig. 18   Open outlet electric water heater   42
Fig. 19   A pillar tap   46
Fig. 20   All-plastic *Opella* tap with shrouded head   47
Fig. 21   Drain-cock with hose connector outlet   52

Fig. 22	Screw-down stop-cock	53
Fig. 23	A *Markfram* mini stop-cock	54
Fig. 24	A gate valve	56
Fig. 25	*Portsmouth* pattern ball valve	58
Fig. 26	An equilibrium ball valve	61
Fig. 27	Modern diaphragm ball valve by Pegler-Hattersley Ltd	63
Fig. 28	*Burlington* pattern flushing cistern	66
Fig. 29	A direct action flushing cistern	68
Fig. 30	Siphoning mechanism from direct action flushing cistern	69
Fig. 31	Wash-down lavatory pan	70
Fig. 32	Single trap siphonic lavatory suite	72
Fig. 33	Double trap siphonic lavatory suite	73
Fig. 34	A magnesium anode in a cold water storage cistern	84
Fig. 35	The effects of internal corrosion on radiators	88
Fig. 36	A lagged galvanised steel cold water storage cistern	93
Fig. 37	A lagged round polythene storage cistern	93
Fig. 38	A cylinder hot water system forms an open-ended U tube	101
Fig. 39	A spring safety valve	102
Fig. 40	Traps	105
Fig. 41	Two-pipe drainage systems	108
Fig. 42	Two-pipe drainage from a multi-storey building	109
Fig. 43	One-pipe drainage from a multi-storey building	111
Fig. 44	Single stack drainage	113
Fig. 45	The *Barvac* antisiphon trap	114
Fig. 46	Two ways of connecting the bath waste to a single stack	114
Fig. 47	A uPVC rain water system	121
Fig. 48	Detail of fixing an OSMA roundline uPVC gutter	123
Fig. 49	Typical pre-war suburban drainage system	127
Fig. 50	Drain inspection chamber with intercepting trap and fresh air inlet	128
Fig. 51	An OSMA drain GRP inspection chamber	131
Fig. 52	'Private sewers' and 'public sewers'	133
Fig. 53	Clearing a blocked intercepting trap	135

Fig. 54	Clearing a blocked drain	136
Fig. 55	Deep and shallow wells	141
Fig. 56	Typical cesspool	145
Fig. 57	A septic tank	148
Fig. 58	Septic tank with filter	150
Fig. 59	Land drainage trenches	152
Fig. 60	Subsoil drainage	155
Fig. 61	Hedges and ditches	156
Fig. 62	A few examples from the *Typay* range of non-manipulative compression fittings	161
Fig. 63	An *Acorn* push-fit pipe connector	163
Fig. 64	Making a manipulative compression joint	164
Fig. 65	Making an integral ring soldered capillary joint	166
Fig. 66	U-can copperbend	170
Fig. 67	Making a wiped soldered joint	174
Fig. 68	A wiped lead-to-copper or lead-to-iron connection	177
Fig. 69	Making a *Staern* or soldered spigot joint	178
Fig. 70	A finger-wiped joint	179
Fig. 71	Making a solvent weld joint with Marley pvc tube	182
Fig. 72	Making a ring-seal joint	183
Fig. 73	*Kontite* compression coupling used with polythene tubing	186
Fig. 74	Snap ring connection of pitch fibre pipes	187
Fig. 75	A modern sink unit	190
Fig. 76	Fitting a pillar tap into a stainless steel sink top	191
Fig. 77	Sink waste components	192
Fig. 78	Swivel tap connector with compression inlet	194
Fig. 79	A *Kontite* tap extension piece	195
Fig. 80	Plumbing in a washing machine	199
Fig. 81	The *Kontite* 'Thru-flow' valve	201
Fig. 82	Fitting an outside tap	205
Fig. 83	An extended garden water supply	207
Fig. 84	Plumbing in a water softener	208
Fig. 85	Frames for acrylic plastic baths	213
Fig. 86	The connections to a modern bath	215
Fig. 87	At the foot of an old bath	217
Fig. 88	Shower design requirements	219

Fig. 89   *Deltaflow* basin mixer with pop-up waste   224
Fig. 90   Using a cranked bath or basin spanner   226
Fig. 91   'Bridging the gap' between water supply pipe and
            tap tail.   228
Fig. 92   'Over-rim' and 'rim supply' bidets   229
Fig. 93   *Deltaflow* bidet set with ascending spray and
            pop-up waste   230
Fig. 94   Design requirements for rim supply bidet with
            ascending spray   231
Fig. 95   Removing an old lavatory pan with cemented-in S
            trap outlet   233
Fig. 96   *Multikwik* drain connectors   235

Introduction

The plumber's trade and, indeed, the very meaning of the word plumbing, have changed radically during the past four decades.

Before the Second World War a plumber was primarily a 'worker in lead and zinc'. Contemporary plumbing manuals were largely concerned with the properties and use of these metals in domestic water services and 'above ground' drainage, as well as in the provision of valley gutters and other roof work, and in the provision of damp proof courses in buildings.

Plumbing has now come to mean – certainly to the general public and probably to the majority of plumbers – the provision of hot and cold water services and the installation and maintenance of sanitary fittings and of domestic equipment like automatic washing machines and dish washers. It has also come to mean both 'above ground' and 'underground' drainage work.

Lead, because of its price and the hazards of lead poisoning, is now never used in new water supply and drainage installations, although the professional plumber still needs skill in handling it for replacement and repair work in older properties.

The materials which the modern plumber uses are copper, stainless steel and a wide variety of plastics; and it is on plumbing in this modern sense, and the safe and effective use of these materials, that this book concentrates.

The general availability of these materials and the ease with which many of them can be handled has brought a great deal of plumbing work, previously exclusive to the professional, within the scope of the d-i-y enthusiast. I do not believe that any householder can do any plumbing job. Neither do I believe that no one but an experienced professional should ever touch any part of the domestic plumbing system. As with household painting and decorating there is ample scope within the range of plumbing activities for both the handyman and the professional. Such basic maintenance jobs as replacing washers, unblocking choked waste pipes and renewing ball valves certainly do not demand a long apprenticeship.

As for more ambitious projects – such as installing sink units, plumbing in washing machines and replacing wash basins and baths – the householder who understands his plumbing system will be in a better position to know whether these are jobs that he can tackle himself or whether he should call in a professional.

With this in mind the early chapters of this book deal with plumbing principles: the design of hot and cold water and drainage systems and the uses, protection and maintenance of plumbing equipment. The plumber – whether amateur or professional – who has mastered these principles can then practise his plumbing techniques on the projects outlined in later chapters.

In writing this book I have tried to keep up with the latest developments while never allowing myself to forget that the plumbing systems in *most* British homes were installed two decades or more ago.

Metrication

The somewhat erratic progress of metrication presents problems to anyone writing a comprehensive manual of domestic plumbing.

A lot of plumbing equipment is now sold in metric sizes. Some – at the time of writing – is not. Taps, for instance, are still sold as $\frac{1}{2}$ in (for sinks and basins) or $\frac{3}{4}$ in (for baths) though there is a tendency now to refer to them as 15 mm or 22 mm respectively. These sizes are not the dimensions of the taps, but are the metric dimensions of the *copper* tubing to which they will probably be connected.

Where appropriate I have attempted to give both the metric dimension and its Imperial equivalent. Please note though that this equivalent is not necessarily a straight translation of the metric size.

The metric equivalents of thin-walled $\frac{1}{2}$ in, $\frac{3}{4}$ in and 1 in copper and stainless steel tube are 15 mm, 22 mm and 28 mm. The apparent discrepancy arises because the Imperial measurement is of the tube's inside diameter. The metric measurement is of its outside diameter.

To add to the confusion, thick-walled lead and iron pipes are still designated by their internal dimension – now translated into metric units. Thus, the metric equivalents of $\frac{1}{2}$ in, $\frac{3}{4}$ in and 1 in lead pipe are 12 mm, 20 mm, and 25 mm respectively!

When describing water services, I have assumed the use of thin-walled tube and have therefore given the appropriate dimensions as 15 mm ($\frac{1}{2}$ in), 22 mm ($\frac{3}{4}$ in) or 28 mm (1 in).

For approximations, I have used Imperial units only. I think that for some years to come, 'about a couple of inches' or 'about half a gallon', is likely to be more meaningful to the British reader than, 'approximately 50 mm', or 'about $2\frac{1}{4}$ litres'.

1

The Plumber's Tools

Childhood memories of a plumber arriving with an overflowing tool bag – and always the one tool essential for the
particular job left behind! – may deter many home handy-
men from attempting plumbing work. Times have changed.
The owner of a modern – or modernised – house with copper
plumbing will probably be surprised to find how few tools
he needs to carry out repair and maintenance work or even
to engage in fairly ambitious improvements to the existing
plumbing system. Much of the necessary equipment he will
already have by him for other household tasks.

A sturdy *ladder*, preferably a purpose-made loft ladder,
for getting up into the roof space is a *must* – and if you
called in a professional plumber he would expect you to
provide one. You will also need a reliable *electric torch*.
There may be a lighting point in your roof space but it is
unlikely that it will illuminate every dark corner in which
parts of the plumbing installation are to be found. Choose
a torch that can be set down without risk of its rolling away
– and renew the battery directly it begins to dim.

You'll need a couple of good *adjustable wrenches* and one
of them should have a rather wider grip than the wrench
that you get with a car maintenance kit. It may be required

to turn that large nut immediately beneath the lavatory flushing cistern.

A set of assorted *spanners* is essential. So are *screwdrivers*, a couple of pairs of *pliers*, a *hacksaw* and a *metal file*. A work bench with a *vice* can be very useful at times but it could hardly be regarded as essential.

Other useful equipment includes a *power drill* with variable speeds for wire brushing, hole cutting and drilling walls and floors for fixing screws, a *blow torch*, *bending springs* for 15 mm and 22 mm copper tubing and a *wheel tube cutter* with a reamer. There is really no need to get these last two items though unless you have a major plumbing project in mind. Even then, you can get by without them.

There are, of course, other tools that you may need very occasionally – an *immersion heater spanner* if, for instance, you want to fit or renew an immersion heater in your hot water storage cylinder. A bath or *basin spanner* or an adjustable *tap wrench* can be invaluable for unscrewing the very tricky back nuts that hold taps in position. Unless you are going into plumbing professionally it is not worth your while to buy specialist tools like these. Borrow or hire them as the need arises.

The only essential drainage tool that every householder should possess is a *force cup* or sink waste plunger. The use of these cheap and simple gadgets is described on p. 116. Drain rods are, again, tools for the professional. If your drains block sufficiently frequently to justify the purchase of drain rods, then there is something wrong that demands more radical attention.

Apart from tools there are a number of items of equipment that no home plumber can afford to be without – an aerosol can of *penetrating oil* for freeing corroded nuts, a jar of *petroleum jelly* (Vaseline), some *solder wire* and *flux* for use with soldered capillary joints, a roll of *waterproof building tape*, a roll of *PTFE plastic thread sealing tape*, a tin of *jointing compound*, a tin of *non-setting mastic filler* such as *Plumbers Mait* and an *epoxy resin repair kit*.

The use of all these tools and pieces of equipment is fully described elsewhere in this book.

Where can they be bought? Most of them at virtually any d-i-y shop or departmental store with a large d-i-y section. If you have trouble obtaining any particular piece of equipment look up 'Builders Merchants' in the yellow pages of your local telephone directory. Most, if not all, builders merchants are as happy to supply individual householders as they are building firms.

One final point – don't economise when buying tools. Generally speaking you get what you pay for in this field. Go for tools of a known brand that the shopkeeper can recommend as being reliable. If one make is materially cheaper than another the chances are that it will prove to be a poor buy in the long run.

2

Domestic Cold Water Supply

The Water Authority's stop-cock

Let's start at the beginning – with the water supply to your own home.

Near the front boundary of your property, possibly set into the surface of the public footpath just outside the front gate, you'll find a small hinged metal cover. Raise it and you'll see a stoneware *guard pipe*, at the base of which, about 3 feet below ground level, will be the Water Authority's *stop-cock*. Your responsibility, as a householder, for the water installation of your home begins at this point.

The stop-cock may have an ordinary tap handle or it may have a special shank that can only be turned with the Water Authority's turn-key. Most modern homes will have another stop-cock *inside* the householder's home. Where the Authority's stop-cock provides the sole means of cutting off the domestic water supply, the occupier should own or have access to an appropriate turn-key. Whenever you need access to this stop-cock, *do not forget to replace the cover if you leave it, even if for 'only a few moments'*. Particularly when situated in the public footpath a raised stop-cock cover can present a serious hazard to pedestrians and could make

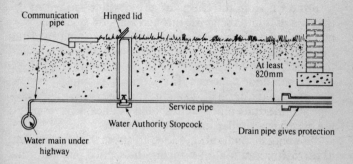

Fig. 1   Water authority stop-cock and service pipe

you liable for very substantial damages if anyone were injured as a result of your neglect.

The service pipe

The length of pipe connecting the water main to the Water Authority's stop-cock is called the *communication pipe*. From the stop-cock to the house this water supply pipe is referred to as the *service pipe*. Before the Second World War, service pipes were almost always made of lead. Nowadays they are likely to be of copper though polythene tubing is sometimes used.

Of course, you cannot see the underground service pipe unless you go excavating! It should rise slightly towards the house to ensure that any air bubbles can escape freely. As a frost precaution, though, it is important that it should be at least 80 cm (2 ft 6 in) below the surface of the ground for the whole of its length. Subsequent garden landscaping or the construction of a drainage channel round the walls of the house may have reduced this soil cover, so do check as it increases the risk of frost damage.

Where the service pipe passes under the footings of the house it should be threaded through lengths of drain pipe to protect it from the effects of any possible settlement. If the pipe rises into the house through a hollow boarded floor make sure it is protected from icy draughts that may whistle through the underfloor space during the winter. It is usual to thread it through a length of 15 cm (6 in) drain pipe set on the sub-floor concrete. The service pipe must pass up the centre of the drain pipe and the space between the service pipe and the drain pipe's walls should be filled with vermiculite chips or a similar insulating material.

Plumbing manuals always used to stress the importance, as a frost precaution, of bringing the service pipe into the home against an internal, rather than an external, wall. This advice was all too frequently ignored. Fortunately modern methods of thermal insulation and the recent practice of cavity wall infilling, have made this precaution less important than it once was.

In most cases the service pipe will enter the house through the kitchen floor, near the kitchen sink. From the point of its emergence it is often referred to as the *rising main*.

The rising main

The householder's main stop-cock should be fitted into the rising main a few inches above floor level. Immediately above it there should be a *drain-cock*. These two fittings allow the water supply to be cut off and the rising main drained when required. Combined stop-cock/drain-cocks are available.

The importance of the main stop-cock cannot be overstressed. Whether you have a burst pipe, a leaking cold water storage cistern or a jammed ball valve – in fact almost any plumbing emergency – the first step to take is to turn the

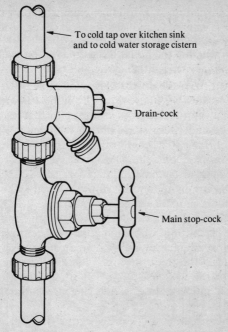

To cold tap over kitchen sink
and to cold water storage cistern

Drain-cock

Main stop-cock

Fig. 2   Householder's main stop-cock and drain-cock

water off. Every member of the household should be familiar with the stop-cock's position and purpose.

At least one *branch water supply pipe* will be taken off the rising main in the kitchen. This is the 15 mm ($\frac{1}{2}$ in) supply to the cold tap over the kitchen sink. This tap provides the household's drinking and cooking water. It *must* be supplied with water direct from the rising main, not from a cold water storage cistern which is unsuitable for drinking water.

Other branch supply pipes that might be taken from the rising main include one to an outside tap and one to an automatic washing machine or dish washer.

Direct and indirect cold water services

The course of the rising main, after leaving the kitchen, will depend on whether the house has a *direct* or an *indirect cold water system*. With a direct system all cold water draw-off points – the W.C. flushing cistern, the bath cold tap and the wash basin cold tap – are taken direct from the rising main.

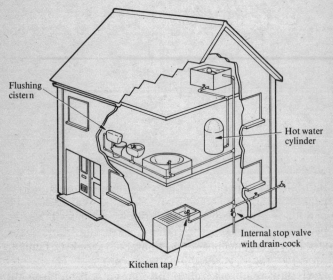

Fig. 3   Direct cold water system

An indirect cold water system is one in which bathroom draw-off points – the cold water taps of the bath and the basin and the lavatory flushing cistern – are supplied from a main cold water storage cistern.

Many Water Authorities, particularly in southern and eastern England, insist on the provision of indirect cold water systems in new buildings, as their water mains are

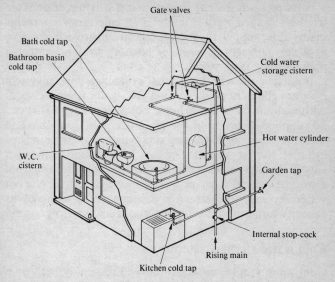

Gate valves

Bath cold tap

Bathroom basin cold tap

Cold water storage cistern

Hot water cylinder

W.C. cistern

Garden tap

Internal stop-cock

Rising main

Kitchen cold tap

Fig. 4 Indirect cold water system

incapable of providing a direct supply during periods of peak demand. An indirect cold water system offsets this period of peak demand by acting as a kind of 'buffer' between the consumer and the water main. Cold water storage cisterns are full when the heaviest domestic demand for water begins at about 7.00 a.m. each day. Their water levels fall as householders visit the bathroom, flush the lavatory cistern and draw off water for baths, washing and shaving. The cisterns then refill slowly during the peak period but more rapidly as demand levels off and pressure in the main increases later in the day.

As far as the householder is concerned, an indirect cold water system makes possible the provision of bath and basin mixers and conventional shower installations. Lavatory cisterns supplied from a main storage cistern fill less noisily and are less prone to condensation too; while the effects of a burst pipe will be less devastating if it is under pressure from a storage cistern. Finally, if an emergency interrupts mains water supply, the storage cistern will provide a reserve of water sufficient at least for lavatory flushing until the mains supply is resumed.

It must be stressed though that a direct cold water system does not eliminate the need for a cold water storage cistern. Although there are whole-house hot water systems that can be supplied direct from the main, the most popular and versatile means of hot water supply is one of the kinds of cylinder storage system. Systems of this kind demand a supply of water under constant, relatively low pressure from a storage cistern. While providing such a cistern for the hot water services it seems only common sense to make its capacity sufficient to supply bathroom cold services as well.

Cold water storage cisterns

Cold water storage cisterns, intended to supply bathroom cold taps and a W.C. flushing cistern, as well as a cylinder storage system, should have an actual capacity of 227 litres (about 50 gallons). *Actual capacity* is the capacity to the invert of the overflow or warning pipe – about 115 mm ($4\frac{1}{2}$ in) from the rim. The amount of water that the cistern would hold if it were filled to the brim is its *nominal capacity*.

Where should the cold water storage cistern be situated? The roof space is the traditional site but an alternative is the upper part of a bedroom or bathroom airing cupboard. If this latter is chosen it is possible to keep all the plumbing out of the roof space, and it will consequently be far less vulnerable to frost damage.

But the roof space in my opinion is still preferable. In an airing cupboard condensation on the cold surface of the cistern can present problems and plumbing noises are all too audible. Furthermore the lower position will mean a reduction of pressure to the bathroom taps and a slower refill of the lavatory flushing cistern; and it will make it impossible to provide a shower on the same floor as the storage cistern without the additional installation of an expensive shower pump.

The risk of frost damage to the plumbing in the roof space cannot be ignored though, and protection is discussed in some detail in Chapter 7. Here it is sufficient to say that pipe-runs within the roof space should be kept as short as possible and, if a flue in regular use passes through the roof space, the cistern should be sited near it. To obtain maximum support for the cistern and its contents (remember that 50 gallons of water weigh nearly a quarter of a ton) the cistern should be placed over one of the dividing walls of the house.

Cold water storage cisterns are made of galvanised mild steel, asbestos cement or of a variety of plastics. Galvanised mild steel is the traditional material and many thousands are in use today. Its disadvantages are its weight and liability to corrosion.

Asbestos cement cisterns cannot, of course, corrode. They are rather heavy though, and have a tendency to absorb some of the water that they contain. They are also prone to accidental damage in storage or installation.

Plastic storage cisterns

The best choice for a modern installation is a plastic cistern. These are available in rectangular form – perhaps reinforced with glass fibre for extra strength – or they may be round and flexible. They cannot corrode, have rounded, easily cleaned internal angles and, above all, are easily handled and sufficiently light for one-man installation.

Before installing a plastic cistern of any kind do read the installation instructions supplied by the manufacturer.

One essential requirement common to all plastic cisterns is that they should stand on a flat, level base – not just on the roof timbers. This can be met by nailing (or 'spiking') some lengths of floor board or a rectangle of chip-board or marine ply to the joists. It is also important that pipe

connections should be made squarely to the cistern walls
so that the plastic material is not under strain.

Holes for pipes are best cut with a hole-cutting attachment
fitted into a brace or drill. Turn the cistern on its side and
fix a piece of wood below the cistern wall to provide support
at the point where the hole is being cut. If you don't have
a hole-cutting tool you can mark out the circumference of
the hole to be cut, drill a series of small holes round the
inside of this circumference, then knock out the piece in the
middle and neaten off with a half-round file.

Manufacturers' advice varies about pipe connections.
Most suggest that two large sized washers, one metal and
one plastic, should be used between the back-nuts and the
cistern walls – the plastic washer being in direct contact with
the wall. They also urge that no boss white or similar jointing
compound should be allowed to come into contact with the
plastic walls. *PTFE plastic thread sealing tape* should be used
where it is necessary to make a screwed connector watertight. However one manufacturer of plastic/glass fibre cisterns advises that plastic washers should not be used in
contact with the cistern walls and says that any water-
proofing compound may be used with his particular cisterns.

One important point that is rarely stressed by manufacturers is the importance of ensuring that the rising main
is securely fixed to the roof timbers and is not supported
solely by the walls of the cistern. This is particularly
important where a galvanised steel cistern is replaced by a
flexible plastic one. Galvanised steel cisterns offer good
support to the rising main. Plastic cisterns – particularly
round flexible ones – do not. Intolerable noise and vibration
may result if you neglect this.

Cistern dimensions

227-litre capacity cisterns vary considerably in dimensions
according to make. A galvanised steel cistern of that
capacity might well be 90 cm (36 in) long by 60 cm
(24 in) wide by 58 cm (23 in) deep. A rectangular plastic

cistern could be 85 cm (34 in) long, 77 cm (30 in) wide and 60 cm (23 in) deep. A round flexible cistern could have a top diameter of 85 cm (34 in), a base diameter of 75 cm (29½ in) and a height of 60 cm (24 in). As problems sometimes arise getting a new cistern through a small trap door into the roof space, it's worth remembering that round, polythene cisterns can usually be flexed through an opening measuring only 60 × 60 cm (24 × 24 in).

Where it is quite impossible to pass a big enough cistern through the trap door the problem can be overcome by linking two smaller cisterns together with a 28 mm (1 in) tube fitted about two inches from their bases. Do not take the distributing pipes to the bathroom cold services and the hot water system from the cistern to which the rising main is connected. This will ensure a continuous flow of water through the two cisterns and will eliminate the risk of stagnation occurring.

Connections to the storage cistern

There are likely to be at least four – possibly more – pipes connected to the cold water storage cistern in an indirect system.

The cold water inlet will be via a *ball valve* fitted about an inch below the cistern rim. Below the level of the *ball valve inlet*, with its lowest point at least an inch above normal water level, is the *overflow* or *warning pipe*. This must be at least 22 mm (¾ in) in diameter and must discharge in the open air in a position where any overflow from it will quickly attract attention.

It used to be the practice to fit a hinged copper flap to the outlet of this pipe to prevent icy draughts blowing up it. Unless regularly lubricated though these flaps usually jammed open or shut and proved ineffective. A simpler, and more effective, method of protection is to extend the overflow pipe inside the cistern and to bend it over so that it dips an inch or so below the surface of the water. The wate then acts as a trap to prevent draughts blowing through tl

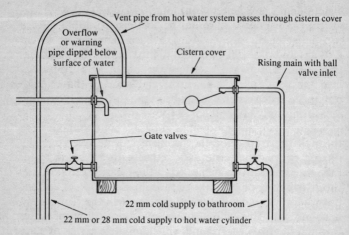

Vent pipe from hot water system passes through cistern cover

Overflow
or warning
pipe dipped below
surface of water

Cistern cover

Rising main with ball
valve inlet

Gate valves

22 mm cold supply to bathroom

22 mm or 28 mm cold supply to hot water cylinder

Fig. 5   The connections to a cold water storage cistern

pipe. Plastic fittings – the Shires *Frostguard* for example –
are available for fitting to the inlets of existing overflow
pipes.

The *cold water distribution pipe* to the bathroom and
lavatory should be taken from a point about two inches from
the base of the cold water storage cistern to reduce the risk
of grit and other debris being drawn into the pipe. In most
suburban homes a 22 mm ($\frac{3}{4}$ in) pipe is big enough. This
will be taken by the most direct route to supply the $\frac{3}{4}$ in
cold tap over the bath. 15 mm ($\frac{1}{2}$ in) branches will be taken
from it to supply the $\frac{1}{2}$ in cold tap over the wash basin and
the lavatory flushing cistern.

Where there is more than one wash basin, or lavatory
suite, a 28 mm (1 in) main distribution pipe may be needed.
It is worth noting at this point that the discharging power
of a 28 mm (1 in) pipe is *twice* that of a 22 mm ($\frac{3}{4}$ in) pipe
and *six times* that of a 15 mm ($\frac{1}{2}$ in) one. Consequently very
few domestic premises are so large that they need a cold
distributing pipe with greater diameter than 28 mm (1 in).

Unless the cold water storage cistern is situated immediately above the bathroom there is likely to be a horizontal length of cold water distributing pipe in the roof space. There may be a further horizontal length of pipe at first floor or ground floor level. Although I say 'horizontal' these lengths of distributing pipe should, in fact, slope down slightly when installed from the storage cistern. This will allow air bubbles to escape and so reduce the risk of air locks forming when the system is drained.

As well as the main cold water distribution pipe, separate 15 mm ($\frac{1}{2}$ in) cold distribution pipes may be taken to supply a shower or a bidet. The reasons why *separate* cold supply pipes may be necessary will be explained when these fittings are discussed. These pipes too should be taken some two inches from the base of the cistern and a slight fall given to all nominally horizontal runs.

The cold water supply pipe to the hot water storage cylinder, also taken from a point two inches from the base of the storage cistern, must be at least 22 mm ($\frac{3}{4}$ in) in diameter. Once again, where the domestic plumbing services comprise more than the minimum of sink, bath and basin a 28 mm (1 in) pipe is desirable. This pipe, which connects to the hot water storage cylinder near to its base, must serve the hot water system *only*. No branch distribution pipe may be taken from it.

Gate valves may be fitted into the cold water distribution pipes so that either the hot or cold services can be isolated if necessary without disrupting the entire plumbing system. Any such gate valves should be fitted very close to the cold water storage cistern and should, under normal circumstances, be kept *fully* open. A water supply pipe is only as wide as its narrowest point.

Protection of the cold water storage cistern

As a protection from frost and contamination, all cold water storage cisterns should be provided with a dust-proof – but not hermetically sealed – cover. They should also be insulated on all four sides and the top – but *not* the bottom – so that they benefit from hot air rising from lower floors. Remember, too, not to lay loft insulation immediately under the tank.

Manufacturers of plastic and asbestos cement cisterns make suitable covers that may be purchased as an 'optional extra'. However a cover for a rectangular cistern can easily be made by cutting a piece of plywood to dimensions an inch or two greater than those of the top of the cistern and by fixing a rim of $\frac{1}{2}$ in × $\frac{1}{2}$ in batten round the edge.

3

Domestic Hot Water Supply

A hot water supply for the whole house can be provided from a multipoint instantaneous gas water heater or by one of the types of cylinder storage hot water system.

Direct cylinder storage hot water systems

Cylinder storage systems are easily installed in both new and older homes and the principles on which they work can be readily understood. They can be operated by any fuel, including electricity, or by a combination of two or more fuels. They may provide hot water only or they may form part of a hot water supply/central heating system.

The simplest form of cylinder storage system – one based on a *direct* cylinder – is illustrated in Figure 6. A closed storage vessel (usually a copper cylinder) with a capacity of about 136 litres (30 gal) is supplied with cold water from the main cold water storage cistern by a 22 mm ($\frac{3}{4}$ in) or 28 mm (1 in) pipe.

A 22 mm ($\frac{3}{4}$ in) vent pipe rises from the top of the cylinder's dome. It is taken upwards and bent over to terminate, open-ended, over the cold water storage cistern. The hot water distribution pipes – 22 mm ($\frac{3}{4}$ in) to the bathroom

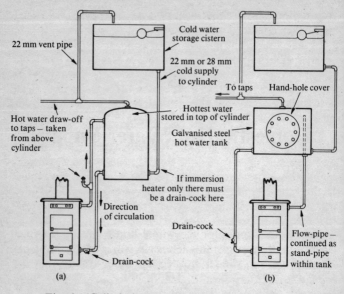

Fig. 6 'Cylinder' storage hot water systems

and 15 mm ($\frac{1}{2}$ in) to the kitchen – are taken from this vent pipe *above the level of the storage cylinder*.

If the water in the cylinder is heated by a boiler a 28 mm (1 in) *flow pipe* will rise from the upper pipe connection (*tapping*) of the boiler to a flow tapping in the upper part of the cylinder wall. A *return pipe* of the same diameter will connect the lower, return tapping of the cylinder to the lower tapping of the boiler.

Lighting the boiler fire ensures a continuous circulation of water between the boiler and the cylinder. As water is heated it expands. The result is that any volume of heated water weighs less than an equal volume of cold. Water heated in the boiler is therefore pushed up the flow pipe into the cylinder by colder, heavier water flowing down the return pipe. This, in turn, is heated, expands and is replaced by further colder, denser water.

In the cylinder the lightest – and consequently the hottest – water 'floats' on top, and is therefore constantly available to be drawn off from the hot taps. Water drawn off is replaced from the main cold water storage cistern. (The cold supply pipe from the cistern connects to the cylinder near to its base.)

Water in an installation of this kind can be heated by an independent boiler in the kitchen, or by the back boiler of a room heater in, say, the living room. This is usually supplemented by a thermostatically controlled electric immersion heater screwed into an immersion heater boss in the cylinder dome. During the summer when the boiler fire is let out the immersion heater can be switched on.

When the time comes to replace an immersion heater it is important to remember that only water *above* the level of the electric element will be heated by it. Vertically fitted immersion heaters should extend to within two or three inches of the base of the cylinder.

Hard water scale or fur

In hard water areas (See Chapter 6) direct cylinder hot water systems can be adversely affected by boiler scale or fur.

When hard water is heated to a temperature of 70°C (160°F) and above, carbon dioxide is driven off and dissolved calcium bicarbonate is converted in to insoluble calcium carbonate. This is deposited as scale on the internal surfaces of boilers and flow and return pipes and on the external surfaces of immersion heaters. As boiler scale is a poor conductor of heat, it insulates the water in the boiler from the heat of the boiler fire. The first indication of its presence is that the water in the cylinder takes longer to heat up than it did formerly. Instinctively the householder piles on more fuel or increases the draught. More scale forms and eventually there will be hissing, bubbling or banging noises as the water in the boiler is forced through ever narrowing channels.

As well as insulating the water in the boiler from the boiler fire, scale also insulates the metal of the boiler from the cooling effect of the circulating water. This can have an even more serious effect. The metal burns away and eventually a leak will develop. Similarly an immersion heater affected by scale will overheat and will rapidly burn out and fail.

Scale prevention

It is possible to introduce acid-based chemicals into a direct hot water system to dissolve scale that has already formed. However, by far the best course of action is to prevent scale from forming in the first place. There are several ways in which this may be done.

1   *Control the temperature of the water.* Scale begins to form only when the water temperature exceeds 60°C (140°F). In soft water areas it is usual to set immersion heater thermostats at 70°C (160°F). In hard water areas the thermostat should always be set at 60°C (140°F).

   You can check the setting of your thermostat by switching off the heater, removing the cover of the heater and inspecting – and if necessary adjusting – the screw that controls the regulator.

   Although it may be possible to control a gas- or oil-fired boiler sufficiently to maintain water temperature at 60°C (140°F) this is rarely possible with a solid fuel boiler.

2   *The provision of a mains water softener* (See Chapter 6). This will eliminate all risk of scale formation but is not everybody's choice.

3   *'Micromet' crystals.* Certain phosphates of sodium and calcium, sold commercially as *Micromet*, stabilise the chemicals that produce hardness so that scale does not form when the water is heated. *Micromet* crystals are placed in a purpose-made plastic mesh container and

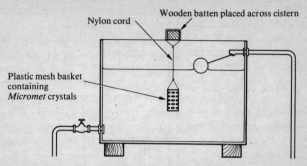

Fig. 7   *Micromet* crystals suspended in cold water storage cistern

are suspended in the water of the cold water storage
cistern. The soluble crystals usually need to be replaced
at six-monthly intervals.

4   Finally – and this is the most radical solution to the
problem of scale formation – the householder may provide an *indirect* instead of a direct cylinder storage hot
water system.

Indirect hot water systems

In an indirect hot water system the water stored in the hot
water cylinder does not circulate through the boiler. It is
heated indirectly by a closed coil or heat exchanger, within
the cylinder, to which the flow and return pipes from the
boiler are connected.

This closed circulation between the boiler and cylinder is
known as the *primary circuit*. It is supplied with water from
its own small (say 23 litre/5 gal) *feed and expansion tank*
and has its own vent pipe which terminates open-ended over
this tank.

Within the primary circuit the same water circulates over
and over again. It cannot be drawn off from the taps and
only the very tiny losses from evaporation need to be made
up from its own feed and expansion tank. When the system

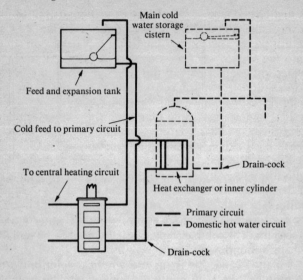

Main cold
water storage
cistern

Feed and expansion tank

Cold feed to primary circuit

To central heating circuit

Drain-cock

Heat exchanger or inner cylinder

——— Primary circuit
– – – Domestic hot water circuit

Drain-cock

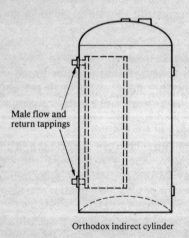

Male flow and
return tappings

Orthodox indirect cylinder

Fig. 8   An indirect hot water system

is first heated hard water chemicals precipitate out to form a very small amount of scale within the boiler and flow and return pipes. Thereafter – unless fresh water is introduced into the system – no more scale will form. Although the water in the outer part of the cylinder, will be hot enough for all domestic purposes, it will rarely reach the high temperatures at which scale is produced.

It is important to note that when the primary circuit is cold there should be only two or three inches of water in the feed and expansion tank. This is because when the system is heated, the water within it will expand and flow back into this tank rising to above the level of the ball float, and if the tank is too full when the system is cold this expansion will result in water overflowing through the tank's overflow or warning pipe. When the water cools again and contracts, fresh, hard, water will flow into the tank to replace that which has overflowed.

Indirect hot water systems also enjoy a relative 'immunity' to internal corrosion. Dissolved air, a prerequisite of corrosion, is driven off from the water in the primary circuit when the system is first heated.

For this reason, an indirect hot water system should always be provided when even the smallest central heating system is installed with hot water supply. Even if hot water supply only is required an indirect system is recommended in areas where the water supply is either hard or corrosive.

Self-priming indirect cylinders

Self-priming indirect cylinders do not have a separate feed and expansion tank. They have specially designed 'inner cylinders' which also serve as heat exchangers.

The cylinder fills from the main cold water storage cistern. As it does so water overflows, via the inner cylinder, into the primary circuit. An air bubble then forms within the inner cylinder to prevent the return of the primary water. The inner cylinder is designed to accommodate the expansion of the water in the primary circuit when it is heated.

Self-priming indirect cylinders provide a cheap and simple way of converting a direct hot water system to an indirect one. But it is generally agreed that they give a less positive separation of the primary and domestic hot water than that found in a conventional indirect system.

It is very important that water in the primary circuit of a self-priming system should never reach boiling point. If it does, the air bubble will be blown out of the system and the two waters will mix. Where central heating is provided in conjunction with hot water supply it is also important to ensure that the particular model of self priming cylinder that is installed is large enough to accommodate the expansion of the primary and radiator circuits.

Manufacturers include larger-than-standard models in their range for this purpose.

Packaged plumbing systems

Packaged plumbing systems, or 'two in one' hot water supply units, have been developed as a result of the need to provide quickly installed hot water systems in older houses that were previously without a hot water supply, and in self-contained flats created by the conversion of large old houses.

A packaged plumbing system has the same basic design as a conventional cylinder storage system but the cold water storage cistern and hot water cylinder are brought close together to form one compact unit.

Some packaged plumbing systems are suitable for use only in areas where a direct cold water system (see Chapter 2) is permitted. A small cylindrical copper storage cistern of about 45 litres (10 gal) capacity surmounts a conventional copper hot water cylinder. This unit can supply the hot water storage cylinder only. Water supplies to flushing cisterns and bathroom cold taps must be taken direct from the main.

Other models can provide hot and cold water supply for

the whole house. A standard cold water storage cistern of 227 litre (50 gal) capacity is provided above a 136 litre (30 gal) hot water cylinder. A unit of this kind needs only to be placed in position, on a landing or in an upstairs cupboard, and provided with a heat source to give the same service as a conventional hot and cold water system.

An advantage of packaged plumbing units (apart from the ease with which they can be installed) is the fact that the close proximity of cylinder and cistern gives total protection from frost provided that the boiler fire is kept alight or the immersion heater switched on. A disadvantage is the loss of pressure involved in bringing the storage cistern out of the roof. This will give reduced pressure at hot and cold

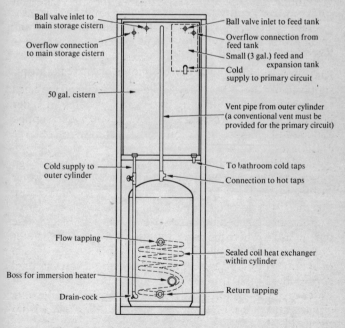

Fig. 9   An indirect packaged plumbing system

water taps, a slower refill of lavatory flushing cisterns, and will make the installation of a shower on the same floor as the unit impossible.

Packaged plumbing units are available with direct, indirect and self-priming indirect cylinders. Units with a conventional indirect cylinder usually incorporate a small feed and expansion tank within the main cold water storage cistern.

Water heating by electricity

An electric immersion heater fitted into a hot water storage cylinder can provide an extremely convenient way either of supplementing a boiler hot water supply or of providing, on its own, an entire hot water system. However unless the system is properly designed, thoroughly insulated and intelligently used it can prove to be extremely expensive to run.

A system of this kind is most economical where bathroom

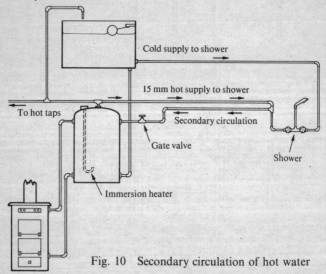

Fig. 10   Secondary circulation of hot water

and kitchen are in close proximity; and this is usually the case in modern house design. This arrangement cuts out long 'dead legs' – lengths of pipe between the storage cylinder and the hot water taps in which expensively heated water will cool after hot water has been drawn off. Ideally the cylinder should be situated closest to the hot water tap in most frequent use – usually the hot tap over the kitchen sink.

Where there is a 'dead leg' longer than twenty feet, say, it may be worth considering providing a small, independent gas or electric instantaneous water heater at that particular draw-off point.

Electrically heated water must never be permitted to *circulate*. There are several ways in which this can occur. For example, to speed up delivery of hot water at a shower or wash basin, a secondary circulation of hot water is some-

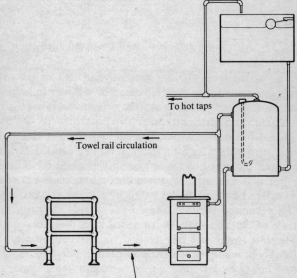

To hot taps

Towel rail circulation

Return taken either direct to a return tapping of the boiler or to the return pipe from cylinder to boiler

Fig. 11 Towel rail circulation connected to direct hot water system

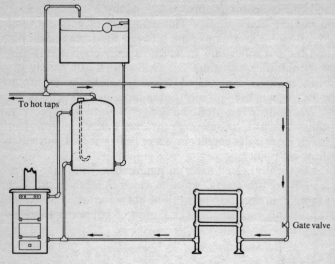

Fig. 12   Alternative arrangement of towel rail circulation

times provided from the hot water cylinder. This may be permissible where the water is heated by a boiler but circulation must be prevented when the immersion heater is in use. A gate valve fitted into the return pipe, turned off when the immersion heater is switched on, will achieve this.

I said earlier that, for even the smallest new central heating system, an indirect hot water system was essential. This is true but it is not unusual for a chromium plated copper towel rail to be connected to a direct hot water system. This can be permitted because a towel rail of this kind will not corrode in the same way as a pressed steel radiator.

Where a towel rail is provided, its circulation should preferably be wholly below the level of the immersion heater. If this cannot be arranged a gate valve can, once again, be provided to prevent circulation when the immersion heater is in use.

Under certain circumstances back circulation through the

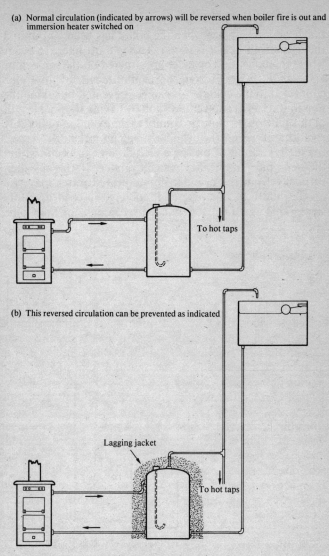

(a) Normal circulation (indicated by arrows) will be reversed when boiler fire is out and immersion heater switched on

To hot taps

(b) This reversed circulation can be prevented as indicated

Lagging jacket

To hot taps

Fig. 13   Reversed circulation – cause and cure

boiler may take place when the boiler has been let out and the immersion heater is in use. This is most likely to occur where the boiler and cylinder are situated on the same floor and where there is a long horizontal length of flow pipe connected to the cylinder. It can be prevented by taking the horizontal run at a lower level and by having the final length of this pipe rise in close proximity to the cylinder, preferably enclosing it within the cylinder's lagging jacket.

Finally, a kind of one-pipe circulation can occur within the vent pipe of the hot water cylinder. If this is taken vertically from the cylinder dome, heated water may rise up the centre of this pipe and, cooling, descend down its sides. This again can be avoided by taking the vent pipe

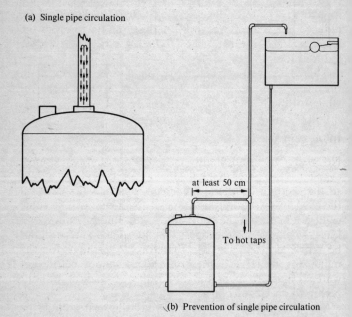

(a) Single pipe circulation

at least 50 cm

To hot taps

(b) Prevention of single pipe circulation

Fig. 14   Single pipe circulation – cause and cure

more-or-less horizontally for about 50 cm (1 ft 6 in) before permitting it to rise vertically to discharge over the cold water storage cistern.

As an indication of the importance of preventing circulation of electrically heated water remember: such water circulating through a 15 mm ($\frac{1}{2}$ in) copper tube at 60°C (140°F) will waste *1.36 units of electricity per foot run per week*.

Lagging

The importance of providing the storage cylinder with an efficient lagging jacket cannot be overemphasised. A 130 litre (30 gal) unlagged copper cylinder maintained at 60°C (140°F) will waste 86 units of electricity per week. Provide it with a 2 in thick lagging jacket and this loss will be reduced to 8.8 units per week. Increase the thickness of the lagging jacket to three inches and weekly heat loss will be reduced to 6 units.

Three inches has, in fact, been found to be the optimum thickness. Increasing it further does not provide a worthwhile saving.

Intelligent use

No matter how thoroughly insulated a hot water storage cylinder is the water in it will slowly lose its heat. As it does so the thermostat of the immersion heater will switch on the element to restore the temperature. There will be some eight hours during the night and – with many families – at least as many hours during the day when hot water will not be required. It is sensible therefore to arrange for the immersion heater to switch on about an hour before hot water is likely to be required and off again about an hour before the need for hot water ends.

When someone is at home all day the heater could be switched on at, say 6.00 a.m. and switched off at about

10.00 p.m. When everyone is out all day the heater might be switched off at about 7.30 a.m. and switched on at about 5.00 p.m.

A programme of this kind can be arranged quite easily by fitting an automatic time switch into the electrical circuit controlling the heater. The programme of such a control device can usually be overridden manually when required – at weekends, say.

Hot water storage cylinders are usually put in the airing cupboard. Never be tempted to strip off the cylinder lagging jacket, or any part of it, to air clothing. You will find that it is more economical to provide a low powered electric airing cupboard heater and to switch that on when required.

Purpose-made electric cylinder water heaters

If electricity is to be the sole means of water heating it is worthwhile considering the installation of a purpose-made electric water heater. These are included in the ranges of all manufacturers of electrical appliances. They are, in effect, very heavily insulated hot water storage cylinders provided with one or more electric immersion heaters. Most require the usual cold water storage cistern as a source of water supply. Some, however, referred to as *cistern heaters*, incorporate their own small cold water storage cistern and are, therefore, identical to the smaller kind of packaged plumbing unit.

Two kinds of electric water heater deserve special mention.

Under Draining Board (or *UDB*) heaters are heavily insulated, rather squat cylinders usually with a capacity of about 90 litres (20 gal) designed for installation under the draining board of the kitchen sink. They are provided with two horizontally fitted immersion heaters. The upper one is intended to be kept permanently switched on (it could, of course, be controlled by a time switch) to maintain a few gallons of hot water for washing and shaving, washing up

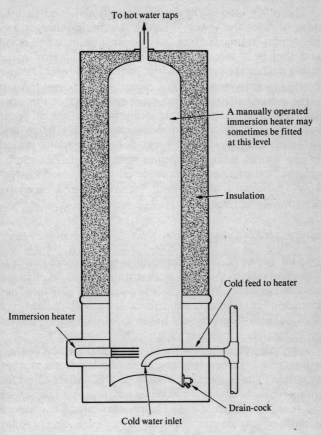

To hot water taps

A manually operated
immersion heater may
sometimes be fitted
at this level

Insulation

Cold feed to heater

Immersion heater

Drain-cock

Cold water inlet

Fig. 15   An off-peak electric water heater

and so on. The other, lower heater, is switched on about an hour before larger quantities of hot water are required for baths and so on.

The other kind of heater is specially designed to take advantage of the lower 'off peak' electricity charges and has a built-in time control. It is heavily insulated and is tall and slim to ensure that the hottest water remains in the upper part of the cylinder and is undisturbed by cold water flowing in at the base. It has a capacity of 227 litres (50 gal) – the estimated daily requirement of hot water for most families. A horizontally fitted immersion heater near to its base switches on when off-peak charge rates begin and switches off again when they end. Some heaters of this kind have a further immersion heater near the top. This can be switched on manually to provide (at the standard charge) a few gallons of hot water if there is an above-average demand for hot water.

Solar water heating

Interest in the possibility of tapping a free source of heat – the radiance of the sun – has grown in recent years as the cost of electricity and of the conventional fossil fuels has increased. In tropical countries a solar installation may provide the sole heat source but in the British Isles it can do no more than supplement some other source.

A typical hot water system, supplemented by solar energy, is illustrated (Figure 16). There are two hot water storage cylinders. The lower one, which may be either direct or indirect, is heated by conventional means. The upper cylinder is an indirect one. The primary circuit consists of a pumped circulation between the cylinder and a solar collecting panel erected on a southern aspect of the roof. Because of its necessarily exposed position an anti-freeze solution must be added to the water in this primary circuit.

The solar panel consists of a system of small bore copper tubes mounted on a black matt base designed to absorb the

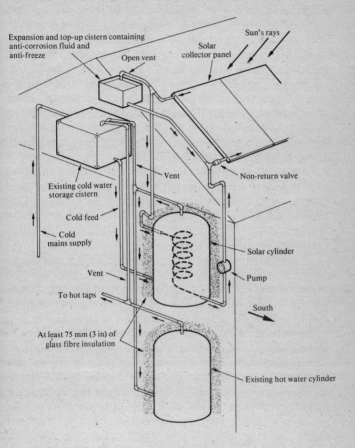

Expansion and top-up cistern containing
anti-corrosion fluid and
anti-freeze

Open vent

Sun's rays

Solar
collector panel

Existing cold water
storage cistern

Vent

Non-return valve

Cold feed

Cold
mains supply

Solar cylinder

Vent

Pump

To hot taps

South

At least 75 mm (3 in) of
glass fibre insulation

Existing hot water cylinder

Fig. 16   A solar heating system

maximum amount of radiant heat. The panel is covered with glass to protect the copper tubing from the cooling effect of the wind and also to add to the heat by producing a 'greenhouse effect'. The short wave-length rays of heat from the sun pass through the glass but the longer wave-length heat rays, reflected from the panel, are trapped by it. The panel should be installed at an angle of 45°, facing south. Mountain climbers will know that the radiant heat of the sun is unaffected by air temperature or by a cooling wind. It is possible to be badly sunburned while climbing among snowdrifts on a mountain top. Thus, even on a cold winter's day, provided that the sun is shining, water passing through the solar panel will be pre-heated.

When hot water is drawn off for domestic use from the vent pipe above the lower hot water cylinder, it is replaced by pre-warmed water from the upper solar cylinder. Less fuel or electricity will be needed to bring the water in the lower cylinder to the required temperature and energy – and money – will be saved.

The value of a solar heater depends to a considerable extent on the geographical position of the property that it serves. A solar heating system on the Isle of Wight or on the outskirts of Bournemouth will prove to be a good deal more effective than one in – for instance – the Lake District.

At present it takes many years for energy savings to balance the very considerable cost of installation. However fuel prices seem destined to rise indefinitely so the gap between the cost of installation and the saving that can be expected is likely to narrow. It may well be that solar water heating will be an important feature of tomorrow's world.

Air locks – their cause and cure

Air locks are perhaps the most common failing in any cylinder storage hot water system.

Typically, while running a bath the householder finds that the flow from the hot water tap diminishes, and perhaps, after much spluttering and bubbling, ceases altogether. After a few minutes lapse the flow may resume. If it does not the chances are that all the hot taps are similarly affected.

An air lock that fails to clear itself can usually be cured by connecting one end of a length of hose to one of the taps giving trouble and the other end to the cold tap over the kitchen sink. First open up the air-locked tap and then the sink tap. Water under mains pressure from the cold tap over the kitchen sink will almost certainly blow the air bubble out of the system.

Always trace the cause of a recurring air lock. Commonly it is an insufficient flow of water from the cold water storage cistern to the hot water storage cylinder.

Under normal circumstances, water level in the cylinder vent pipe is the same as that in the cold water storage cistern. As water level falls in the cistern it falls at an equal rate in the vent pipe. However, if the cold water distribution pipe from the cistern to the cylinder is only 15 mm ($\frac{1}{2}$ in) in diameter instead of 22 mm ($\frac{3}{4}$ in) or 28 mm (1 in), water will not flow from cistern to cylinder quickly enough to replace that drawn off from the bath hot tap. Water level will fall more quickly in the vent pipe than in the cistern, and eventually the level will fall to the horizontal length of distributing pipe supplying the bathroom and air will enter the pipe: an air lock will have been created.

Any obstruction in the cold supply pipe to the cylinder has the same effect. If this is a 22 mm ($\frac{3}{4}$ in) pipe, then any gate-valve fitted into its length must also be 22 mm ($\frac{3}{4}$ in) *and must be kept fully open*. Partial closure has the same effect as reducing the diameter of the pipe. A flake of rust from the storage cistern can also reduce the flow.

Other possible causes of an air lock are: too small a cold water storage cistern; or an inefficient ball valve supplying the cistern with water. To check, watch the level of water in the cistern while someone draws off water from the bath tap. If the cistern is empty before enough water has been drawn off then one or other of these causes is responsible.

Hot water systems should be designed to permit any air that is drawn into a distribution pipe to escape. 'Horizontal' lengths of distribution pipe should, in fact, have a slight fall away from the vent pipe.

Faulty alignment of the flow pipe from boiler to cylinder can also produce a kind of air lock with somewhat alarming symptoms. This pipe should rise, however, slightly, throughout its entire length. If there is a horizontal length – or worse still a length of pipe that falls away from the boiler – dissolved air driven off in bubbles from the heated water will accumulate at that point causing obstruction. Pressure will build up behind the bubble until eventually it will be forced out into the cylinder and up the vent pipe. If it is sufficiently large it will push water standing in the vent pipe out of the open end of this pipe into the storage cistern.

The householder will hear a tremendous bubbling noise followed by the sound of a descending cascade of water in the roof space.

Gas water heating

Gas may be used for both room and water heating and, because of its convenience and relative cheapness, it has become a favourite fuel during the past decade for hot water and central heating systems.

A gas boiler can be used for hot water supply only in conjunction with either a direct or indirect cylinder system. Small gas boilers, known as *gas circulators*, may be installed in direct connection with the cylinder wall.

Gas back boilers, positioned behind a gas fired room heater, are also popular. The boiler and the room heater can operate independently. Very little heat from the boiler reaches the room in which it is situated and a boiler of this kind can therefore eliminate the need for an electric immersion heater for summer use.

Where it is intended to replace a solid fuel boiler with either an oil- or gas-fired one it is important to ensure that the flue has a proper watertight lining. Condensation within the flue from the products of combustion will cause dampness and stain the upstairs chimney breasts otherwise.

The invention of the *balanced flue* has, in fact, eliminated the need for a conventional flue for gas appliances. A balanced flue appliance can be positioned against any outside wall and may be either wall-hung or floor-standing.

The appliance has a combustion chamber sealed off from the room in which it is situated. Air intake and flue outlet are adjacent to each other on the outside wall. They are thus 'balanced'. Wind blowing against the wall will affect air intake and flue outlet equally and the appliance will continue to function normally.

Instantaneous water heaters

The multipoint gas instantaneous water heater is the scarcely recognisable, 'direct descendant' of the hissing, spluttering – and potentially lethal – Edwardian bathroom geyser.

Instantaneous water heaters may be connected either direct to the main or served from a cold water storage cistern. Before choosing this system, you should check that your cistern is high enough to give adequate pressure. The pressure required varies from heater to heater. The Ascot 'Sovereign' multipoint gas water heater for instance, requires a minimum working head of 2 m (6ft 6 in) measured from the level of the water in the cistern *to the highest draw-off point.*

In an instantaneous water heater, water is heated as it flows through a system of small bore copper tubing mounted on a heat exchanger inside the appliance. Turning on the hot water tap turns on the gas jets which are ignited from a small pilot light.

As well as large multipoint instantaneous heaters, there are also small single point ones which can give economical service over a sink or basin. But these cannot be used for supplying a washing machine or dishwasher.

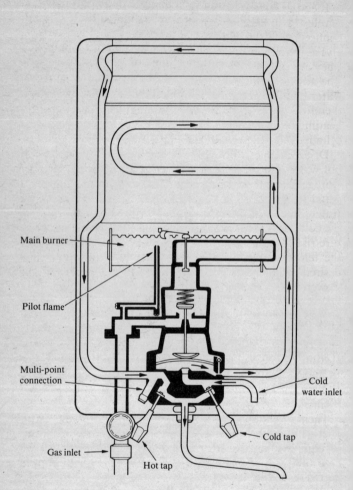

Main burner

Pilot flame

Multi-point connection

Gas inlet

Hot tap

Cold water inlet

Cold tap

Fig. 17 An instantaneous gas water heater

Electric instantaneous water heaters have also been developed in recent years. These are intended for direct connection to the main. Electric water heaters of this kind give a slower flow of heated water than their gas equivalents. They are suitable for spray hand washing or for a shower. The advantage of instantaneous water heaters is that only water that is actually being used is heated. They can therefore prove to be the most economical form of water heating for someone who is out all day, as there is no cylinder full of gradually cooling water.

Disadvantages, though, are that the rate of delivery of hot water is slower than it is from a conventional cylinder storage system and they cannot raise a given volume of water *to* any particular temperature. The temperature of the water at the outlet will depend on the temperature of the cold feed. Most heaters are designed to give a temperature rise of 26°C (110°F). During very cold weather either the temperature of the heated water may be lower than desired or, alternatively, the delivery of hot water may be slower.

Free outlet water heaters

Free outlet water heaters are also intended for direct connection to the water main. They are small, single point appliances that may be operated either by gas or electricity and are used to provide hot water supply at sinks or wash basins.

Their essential feature is a control valve on the water inlet – not on the outlet – of the appliance. When hot water is required the valve is opened. Cold water flows into the base of the appliance and hot, stored water in the upper part is displaced, flowing down a stand-pipe and out of the open outlet.

Free outlet water heaters always used to be fixed above the sink or basin that they supplied. Some modern heaters can now be fitted under the sink or basin and are operated

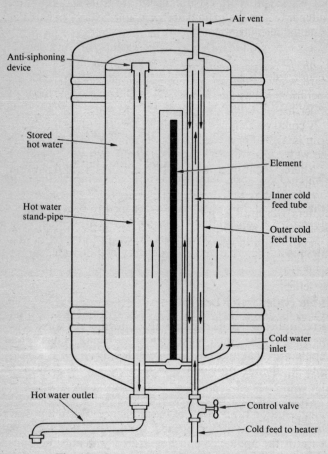

Fig. 18   Open outlet electric water heater

by a valve indistinguishable, at first sight, from an ordinary hot water tap. This valve however controls the inflow of water into the heater and the principle of an open outlet remains unchanged.

Scale in instantaneous and free outlet heaters

Hard water scale can present as serious a problem to the operation of instantaneous and free outlet heaters as to boiler heated hot water systems.

A common failing of free outlet heaters, after they have been installed for some time, is dripping from the outlet as the water heats up. It is often thought that this is caused by a defect in the inlet valve. In fact, it results from expansion of the heated water.

The top of the stand-pipe over which water from these heaters overflows is provided with a small siphoning device, designed to lower the level of the water in the appliance to ½ in (1 cm) or so below its rim after water has been drawn off. This allows space for expansion when the water is heated. Hard water scale can block the siphon and render it ineffective.

The 'cure' is to descale the heater chemically. First, the electricity or gas supply must be cut off. Next disconnect the water inlet to the heater and attach to it a length of rubber tubing. Insert a glass funnel into the other end and raise it above the level of the heater. Slowly pour descaling fluid into the funnel, at about a half pint or so at a time, and stop immediately if there is any sign of foaming back.

After treatment the heater should be thoroughly flushed through before being brought back into use.

Scale prevention

Micromet crystals (see p. 20) can also be used to prevent the formation of hard water scale in instantaneous and free outlet water heaters. The crystals are contained in a purpose made container plumbed into the water supply pipe serving the appliance. A screw-down stop-cock must be fitted on each side of the container so that the crystals can be renewed periodically.

4

Fittings Used in Hot and Cold Water Supply

Taps and 'Supataps'

You are most likely to find *pillar* taps, with a vertical inlet, over your sinks, basins and baths if you have a modern home. *Bib taps*, with a horizontal inlet, may still be found over the sinks of pre-war and immediately post-war houses. But nowadays, the most common use for a bib tap is in providing a garden water supply.

Although there are some all-plastic models, pillar taps are usually made of chromium-plated brass. They come with an easy-clean cover through which the spindle of the tap, surmounted by a *capstan head* (or handle) protrudes. Many modern taps have *shrouded heads*, combining the easy-clean cover and handle in one unit.

However much they may vary in appearance, all such taps work in much the same way. Turning the handle anti-clockwise raises a washered valve (or *jumper*) from a valve seating within the body of the tap letting water pass. Turning clockwise closes the tap. The washer is replaceable.

The one exception in tap design is the *Supatap* (Deltaflow Ltd, Crawley, Sussex). The washer and jumper of a Supatap are fitted into an anti-splash device contained in the nozzle

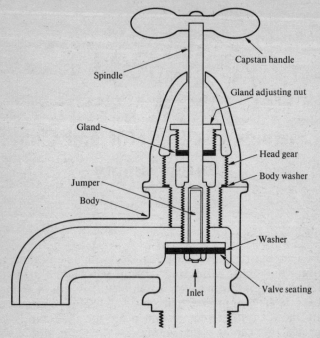

Fig. 19   A pillar tap

of the tap. The tap is turned on or off by turning this nozzle, which is provided with projecting *ears* for this purpose. Early Supataps had metal ears but these became uncomfortably hot when hot water was being drawn off, and now models are provided with insulating *Kemetal* plastic ears.

Apart from their neat appearance and smooth performance a great advantage of the Supatap is the ease with which washers can be changed without cutting off the water. A disadvantage could have arisen when it was wished to connect a hose to the tap outlet. This however has been overcome by the manufacture of special hose connectors with fixing *lugs* to clip over the tap's ears. The body of the

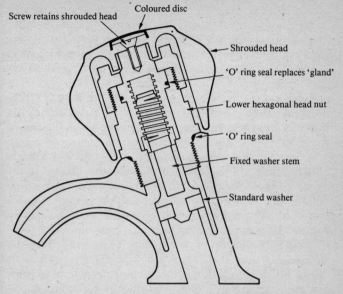

Fig. 20   All-plastic *Opella* tap with shrouded head

hose connector swivels with the tap while the outlet, to which the hose is connected, remains stationary.

Mixers

Bath and basin mixers are two taps with a common spout. The streams of hot and cold water are mixed to the required temperature inside the tap's body by adjusting the two handles. Some bath mixers also have a shower attachment, in which a diversion of the mixed water is made from the bath to the shower head at the flick of a switch.

Conventional bath and basin mixers may be fitted only where there is an indirect cold water supply (see Chapter 2). It is both illegal and impracticable to mix water from the main and water from a storage cistern in any plumbing fitting. With an indirect cold water system and a cylinder storage hot water system both streams of water are under pressure from the same hydraulic head provided by the height of the cold water storage cistern above the appliance.

The cold water supply over the kitchen sink should always come direct from the main so that water used for cooking and drinking is fresh. Sink mixers therefore have a rather different design. The two streams of water pass through the mixer in separate channels, mixing in the air as they leave the spout.

Shower mixing valves may comprise two taps with a single outlet. There are, however, *manual shower mixing valves* which allow the temperature and, with some models, the volume of the water to be regulated by a single control knob. A further development is the *thermostatic mixing valve* which is capable of accommodating minor differences in pressure between the two streams of water. It must be stressed however that, even if it were legally possible to do so, no thermostatic mixing valve is capable of accommodating the great difference in pressure between a mains and a cistern water supply.

Washer changing

You will know that a tap washer needs replacing by a steady, and increasing, drip after the tap has been turned off. Every householder should be able to do this job.

With a conventional tap the first step is to cut off its water supply. The cold supply to the kitchen sink is cut off by turning off the main stop-cock. In most cases you will need to drain the cold water storage cistern before any other tap can be rewashered.

This can be done – without turning off the main stop-cock and thereby stopping the cold supply to the sink – by tying up the float arm of the ball valve supplying the main cold water storage cistern. This will prevent fresh water flowing into the cistern.

If you have an indirect cold water system there is no need to drain off the hot water stored in the cylinder, even if it is the washer of a hot tap that is to be renewed. After tying up the ball valve over the cistern, open up the bathroom *cold* taps and only after these have ceased to run turn on the hot tap to be rewashered. (Since the hot water distribution pipe is taken from the vent pipe above the cylinder only the few pints of water contained in this pipe will have to run to waste.)

Open the tap fully and unscrew the easy-clean cover. You should be able to do this by hand. If you have to use a wrench, pad the jaws to prevent damage to the chromium plated surface.

Raise the easy-clean cover and you will find that you can slip the jaws of your wrench under it to grip the *flats* of the *tap head-gear*. Turn anti-clockwise to unscrew.

After removing the head-gear you may find the valve or jumper, complete with washer, resting on the valve seating. Apply a little penetrating oil to the retaining nut of the jumper and unscrew it, using a spanner. Fit a new washer and re-assemble the tap. Remember that the bath taps ($\frac{3}{4}$ in) take a different size washer from the basin taps ($\frac{1}{2}$ in).

The washer retaining nut can sometimes be very difficult to unscrew. In which case, it's probably best to buy and fit a new washer and jumper complete. You can get these at any d-i-y shop.

Bathroom and kitchen hot taps sometimes have their jumper *pegged* into the tap headgear. The jumper can be turned round and round but cannot be withdrawn. In this event you should make an extra effort to undo the retaining screw. If this just *can't* be done, the trick of the trade is to break the pegging by inserting the blade of a screwdriver between the plate of the jumper and the headgear. Buy a new washer and jumper complete but, before you fit it, burr the stem of the jumper with a rasp so that it fits tightly.

Shrouded head taps are rewashered in exactly the same way as those with capstan handles but the shrouded head must first be removed. The way in which this is done varies with the make of the tap.

With some makes the head can simply be pulled off. With another, the tap must be fully opened and then given a final turn before the head will come off. With yet another there is a tiny retaining *grub-screw* in the side of the head.

The most common method of retention is by means of a retaining screw concealed under the HOT or COLD indicator of the tap. This indicator must be prized off to gain access to it, so try the other methods first!

Supatap washers can be changed without cutting off the water supply. Unscrew and free the retaining nut at the top of the nozzle. Turn the tap on – and keep on turning. Water flow will increase but, just before the nozzle comes off in your hand, it will cease as a *check-valve* falls into position. Tap the nozzle on a hard surface – *not* the glazed surface of a bath or basin! – and turn it upside down. The antisplash device, containing the washer and jumper, will fall out. The washer and jumper can be removed by inserting a coin or the blade of a knife between the jumper plate and the antisplash device. Fit a new washer and reassemble the tap.

When reassembling, remember that the nozzle has a left hand thread, and so has to be turned in the opposite direction from that dictated by instinct.

Faulty valve seatings

Continued dripping after a tap has been rewashered indicates that the valve seating has become scored by grit from the main. Professional plumbers have special tools for grinding the seating to a flat surface again. There are however nylon washer and seating sets that are suitable for d-i-y use.

To fit, unscrew the headgear of the tap and place the nylon lining in position over the existing metal seating. Insert the washer and jumper unit into the headgear. Screw on the headgear and close the tap. This will force the nylon lining into and over the metal seating to give a watertight joint.

Leakage past the spindle

Leakage up the spindle of a tap in use is a fault to which the cold tap over a kitchen sink is particularly prone. Back pressure produced by connecting a garden or washing machine hose can cause this trouble. Alternatively detergent-charged water may have run down the spindle washing the grease out of the gland packing that previously ensured a watertight joint between the spindle and the headgear.

To put this right, you will need to take off the tap's capstan handle. Unscrew the retaining grub-screw in the side of the handle (and put it in a safe place).

The handle should now lift off easily. If it does not, unscrew the easy clean cover and open the tap fully. Place pieces of wood – clothes pegs will do – between the base of the cover and the body of the tap. Turn the handle to close. Upward pressure from the easy clean cover will force off the handle. Remove the easy clean cover.

First try adjusting the gland nut. This is the first nut through which the spindle of the tap passes. Give it half a turn in a clockwise direction. This may well compress the gland packing sufficiently to give a watertight joint.

If the adjustment has already been completely taken up, the gland packing must be renewed. Unscrew and remove the gland nut. Rake out the existing packing and repack with household wool steeped in petroleum jelly (Vaseline). Caulk down hard and reassemble.

In modern taps a rubber 'O' ring seal replaces the traditional gland packing. With such a tap the 'O' ring must be replaced in the event of failure.

Leakage past the spindle should never be ignored. As the gland packing fails the tap will be capable of being turned on and off ever more easily. A rapid cut-off, by 'spinning' the handle with the fingers for instance, can produce the shock-waves of water hammer that can damage pipe lines.

Drain-cocks

Drain-cocks are very simple taps installed to allow the drainage of any parts of the hot and cold water system that cannot be drained from the kitchen or bathroom taps. They are operated by a washered plug, the projecting shank of which can be turned with a spanner. The outlet of the drain-cock has a hose connector.

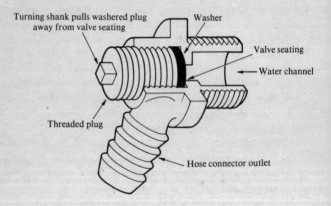

Fig. 21   Drain-cock with hose connector outlet

A drain-cock should be provided immediately above the householder's main stop-cock to enable the rising main to be drained. All boiler operated cylinder storage systems should have a drain-cock fitted into the return pipe from cylinder to boiler at its lowest point – just before connection to the return tapping of the boiler.

Indirect cylinder hot water systems, and direct cylinders heated by electric immersion heaters only, must have a drain-cock fitted at the lowest point of the cold supply pipe to the cylinder. This will enable the cylinder to be drained when required.

Because they are used very rarely, drain-cocks sometimes become blocked by sludge and other debris. If, when the drain-cock is opened, no water flows from it, probe into the outlet with a piece of wire. This should clear any obstruction.

Stop-cocks

Screw-down stop-cocks can be regarded as taps, set into a run of pipe. Some models are chromium-plated and have an easy clean cover and a capstan handle. Most however

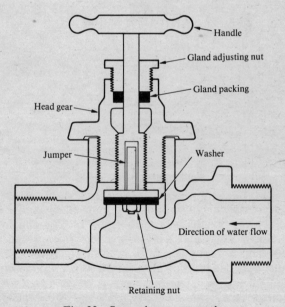

Fig. 22   Screw-down stop-cock

are of unplated brass or gunmetal and have a crutch handle
– like the more basic kind of bib-tap.

When fitting a new stop-cock do make sure it is fitted
the right way round. The arrow engraved on its body must
point in the direction of the flow of water. If the stop-cock
is fitted the wrong way round water pressure will force the
jumper and washer on to the valve seating and no water
will be able to pass, even when it is fully open.

Stop-cocks rarely need rewashering but, when necessary,
it is done in exactly the same way as with a tap. Water supply
must, of course, be cut off first.

Leaking past the spindle is a more common failing and
should receive immediate attention. A constant drip on to

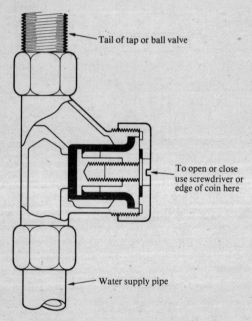

Fig. 23   A *Markfram* mini stop-cock

a wooden floor, particularly if the stop-cock is in a dark and poorly ventilated place can cause dry rot.

Tighten the gland nut or renew the gland packing. Don't forget though that while a tap is normally kept closed a stop-cock is normally left open. Close it before you attempt to remove the gland nut.

Jamming open is perhaps the fault to which screw-down stop-cocks are most prone. To prevent this, open and shut your stop-cocks two or three times, once or twice a year. Before leaving them, open fully and then give a quarter of a turn towards closure. This will not materially reduce water flow but it will make the stop-cock much less likely to jam.

'Mini' Stop-cocks are useful little gadgets designed for fitting between water supply pipes and the inlets to taps or stop-cocks. They can be operated by a screwdriver or the edge of a coin.

Partial closure can reduce flow where pressure is too high. They can also be used to isolate one particular tap or ball-valve for washer changing, for example, without disrupting the rest of the household's plumbing services.

Gate valves

Gate valves look rather like screw-down stop-cocks but they have a larger body and are usually fitted with a wheel, instead of a crutch or capstan handle.

Turning the handle lowers a metal plate or gate, which is normally inside the upper part of the valve body, to close the waterway. They thus give a metal-to-metal seal that is rather less positively watertight than the seal afforded by a stop-cock. They do however give an absolutely unimpeded flow of water when fully open. For these reasons gate valves are used where water pressure is low. Gate valves might be fitted into the cold distribution pipes from the cold water storage cistern to allow either the hot or the cold water services to be isolated. They may also be used to isolate sections of a central heating system.

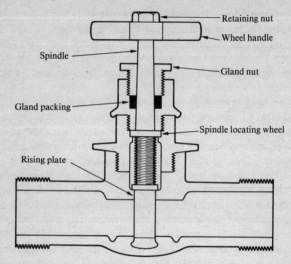

Fig. 24   A gate valve

Very occasional adjustment of the gland screw or renewal of the gland packing is the only maintenance that gate valves are likely to require.

Ball or float valves

Ball valves (sometimes called float valves) are used to maintain the required level of water in cold water storage cisterns, feed and expansion tanks and lavatory flushing cisterns.

All ball valves have a rigid metal or plastic arm at the free end of which is a float (no longer necessarily a 'ball'). As water level in the cistern falls the float falls with it, and the arm activates the valve to allow water to flow in. When the incoming water fills the cistern to the required level the movement of the arm closes the valve.

Croydon and Portsmouth Ball Valves

In both *Croydon* and *Portsmouth* pattern ball valves the valve is closed by a washered metal plug or piston. The plug in *Croydon* valves moves vertically in the valve body. When open, water splashes noisily into the cistern via two channels built into the sides of the body.

Croydon valves are incurably noisy in action and so are rarely found in modern homes. They are, however, very hard wearing and reliable and are often used in cattle troughs and allotment water cisterns where noise doesn't matter.

The plug in a *Portsmouth* pattern valve moves horizontally within the valve body. It is marginally less noisy in operation if only because water flows from it in a single stream from an outlet beneath the body.

Despite the development of more modern ball valves, *Portsmouth* pattern valves are still in common use and may be found supplying the cisterns of new, as well as old, property. Washer failure is their most common defect. This is indicated by a steady drip, or trickle of water from the cistern's overflow or warning pipe.

To renew the washer you must first cut off the water supply to the valve. Some *Portsmouth* valves have a screw-on cap at the end of the valve body, in which case it must be unscrewed and removed. Next withdraw the *split pin* on which the float arm pivots. remove the float arm and place it on one side. Insert the blade of a screwdriver in to the slot at the base of the valve body from which the bent-over end of the float arm was removed and push the plug out of the end of the valve.

The washer is held in place at the end of the plug by a screw-on retaining cap. It may be possible to slip the blade of a screwdriver through the slot in the plug and then turn the screw cap with a pair of pliers. If, the cap won't move, pick the old washer out from under the cap with the point of a pen knife and force the new one under the rim of the cap. Make sure that it rests squarely on its seating.

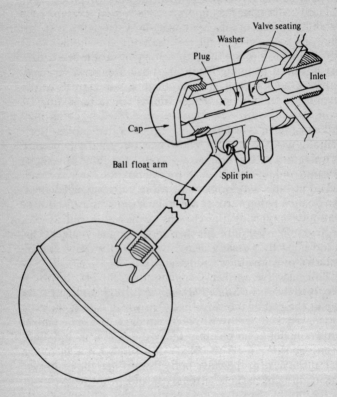

Fig. 25   *Portsmouth* pattern ball valve

Before reassembling the valve, clean the plug and the inside of the valve body with fine abrasive paper and apply a smear of petroleum jelly to the plug.

Hard water scale may make a *Portsmouth* valve jam open or shut. In this event, dismantle the valve and clean the plug and valve body as suggested above. You should renew the split pin on which the float arm pivots at the same time.

Continued dripping from an overflow pipe after a ball valve washer has been renewed could be caused by the water level being set too high, by scoring of the valve seating, or by a 'low pressure' valve having been fitted where a 'high pressure' valve is needed.

To adjust the water level where a *Portsmouth* pattern valve is in use, you will need to bend the float arm. First unscrew and remove the float (which might otherwise break off). Then take the float arm firmly in both hands and bend the float end – upwards to raise the water level, and downwards to lower it.

Reseating tools are used by professional plumbers to reseat ball valves as well as taps. However the home handyman will find it cheaper to replace the valve complete.

Ball valves are classified as high pressure or low pressure according to the diameter of the nozzle orifice. High pressure valves are usually required for cisterns supplied direct from the main, low pressure valves for flushing cisterns supplied from a main storage cistern. Where the storage cistern is only a foot or so above the level of the flushing cistern – as it may be with a packaged plumbing system (see Chapter 3) – a *full-way* valve, giving maximum flow, may be required. Fitting a low pressure valve where a high pressure one is needed will give continual leakage. To fit a high pressure valve where low pressure is needed will result in the cistern refilling unacceptably slowly.

Modern ball valves have detachable nozzles that can be converted from low to high pressure in a matter of minutes. Older valves usually have HP (high pressure) or LP (low pressure) stamped on the valve body.

Ball valve noise

Croydon and *Portsmouth* ball valves may produce unaccept-
able noise as water rushes through them into the cistern.
They may also produce vibration – noticeable perhaps as
a steady humming noise that can be temporarily stopped
by turning on the cold tap over the kitchen sink – and the
heavy reverberations of water hammer.

Vibration results from the movement of the float as it
'bounces' on ripples produced by inflowing water. This
movement is transmitted to the valve and thence to the
rising main which will amplify the sound out of all pro-
portion to its original cause. This is particularly true when
the rising main is made of copper. Water hammer results
from the sudden closure and reopening of the valve as the
ball float bounces on the ripples when the cistern is nearly
full.

It used to be the practice to fit *silencer tubes* into the
outlets of *Portsmouth* pattern ball valves to reduce this ripple
formation. The silencer tube extended from the valve outlet
to a level below that of the water in the cistern. Incoming
water therefore entered below water level and splashing and
ripple formation were eliminated.

Silencer tubes are now banned by Water Authorities who
consider that they produce a risk of back siphonage and
consequent contamination of the mains water if, at any
time, the main is under negative pressure.

A first step to be taken to reduce vibration is to make
sure that the rising main is securely fixed and supported
throughout its length. Fixing clips should be fitted at 1 yard
intervals.

A *stabiliser* can be fitted to the arm of the ball float to
prevent it from bouncing on every ripple. Commercial
stabilisers – a plastic arm which clips to the float arm and
has, at its lower end, a plastic disc – are available. A stabiliser
can be improvised from a small plastic flower pot. Drill
two holes opposite to each other in the rim. Join these with

a loop of copper wire or nylon cord and hang the pot over the float arm so that it is suspended in the water.

Another possible 'cure' is to cut out the final 1 ft 6 in of the copper rising main and replace it with polythene tube.

Finally – and this is likely to be the most certain remedy – the ball valve can be replaced with one designed to give more silent service.

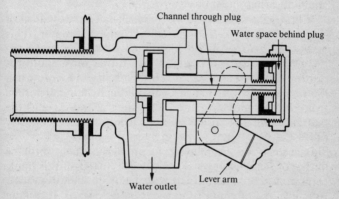

Fig. 26   An equilibrium ball valve

Equilibrium Ball Valves

Equilibrium ball valves were developed principally for use in areas where there is fluctuation in the pressure of water in the main. In such areas water pressure is likely to be high at night and low during the day when there is a heavy demand. Under these circumstances neither a high pressure nor a low pressure valve is likely to give satisfactory service.

A typical equilibrium ball valve looks like a *Portsmouth* valve but it has a channel drilled through the plug to allow water to flow through to a watertight chamber behind it. Thus pressure is equal on each side of the plug. It is possible to have a wide nozzle orifice that will permit rapid filling when pressure is low but which will not result in leakage when pressure is high.

Equilibrium valves can also help eliminate water hammer. Much of the 'bounce' experienced with *Portsmouth* valves is caused by the conflicting forces of the buoyancy of the float and the pressure of water in the main. With an equilibrium valve, since water pressure is equal on both sides of the plug, mains pressure plays no part in forcing the valve open. It is activated *solely* by the movement of the float arm.

Garston (*BRS or diaphragm*) ball valves

Garston, (*BRS* or *diaphragm*) ball valves were developed some years ago at the Government's Building Research Station at Garston in an effort to eliminate noise and other ball valve troubles.

A *Garston* valve is quite different in appearance and action from the traditional *Croydon* and *Portsmouth* valves. The float arm pushes a small metal or plastic plug against a large rubber diaphragm to close the nozzle of the valve. This nozzle is made of hard wearing nylon and, in modern versions, is demountable to permit rapid change from high pressure to low pressure use.

The valve is immune to the effects of scale and corrosion because its only moving part – the small plug – is shielded from the inflowing water by the rubber diaphragm.

Early models of *Garston* valves had a conventional low level outlet and were fitted with a silencer tube. Since these have now been banned, manufacturers have produced valves with an overhead outlet into which a sprinkler device is fitted. This ensures that water enters the cistern in a gentle shower rather than in a rushing stream.

Garston valves may be made of metal or plastic. They are all provided with some means of adjusting water level without having to bend the float arm.

The only faults likely to be encountered with this kind of valve are: the diaphragm jamming against the valve nozzle; and debris from the main – or from a main cold

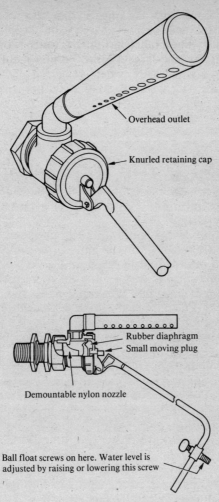

Overhead outlet

Knurled retaining cap

Rubber diaphragm
Small moving plug

Demountable nylon nozzle

Ball float screws on here. Water level is
adjusted by raising or lowering this screw

Fig. 27 Modern diaphragm ball valve by Pegler-Hattersley Ltd

water storage cistern – blocking the space between the nozzle and the diaphragm.

To remedy these faults cut off the water supply to the valve. Unscrew the large knurled retaining nut, free the diaphragm and remove any debris.

5

The Lavatory Suite

In many ways, the lavatory suite is the most important unit of household plumbing equipment. Even the least exacting person will expect it to be readily available, unobtrusive and 100% efficient. A wrongly installed or badly maintained suite can be a positive danger to health.

All flush lavatories comprise a self-cleansing pan with an outlet connected to the drainage system and a flushing cistern (sometimes described in the building trade as *water waste preventer* or WWP) designed to give a measured 9 litre (2 gal) flush.

The Burlington flushing cistern

No one nowadays would install an old-fashioned high-level *Burlington* flushing cistern in a new home. However the tough, hard wearing qualities of these cisterns ensure that there are still many thousands of them in use. They are often to be found in external lavatory compartments where their noisiness is not a serious disadvantage.

Burlington pattern cisterns are made of cast iron and are constructed with a well in the base. A stand-pipe rises from the centre of this well to terminate open-ended an inch or

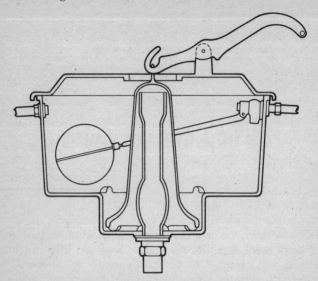

Fig. 28   *Burlington* pattern flushing cistern

two above full water level. The lower end of the stand-pipe connects to the flush-pipe.

A heavy iron bell, with lugs at its base to permit water to pass under the rim, rests in the well, covering the stand-pipe.

To flush the cistern the bell is raised – usually by means of a chain – and is then suddenly released. As it falls heavily into the well its wedge or conical shape forces water trapped within it through the stand-pipe into the flush-pipe.

The descending water takes air with it, creating a partial vacuum. The remaining water in the cistern is then sucked up under the rim of the bell and down the stand-pipe to flush the lavatory pan. This siphonic action continues until the cistern has emptied sufficiently to permit air to pass under the rim of the bell.

Failure of a cistern of this kind to flush is almost always

due to the water level in the cistern being too low. This can be corrected by bending the arm of the ball valve (see p. 59). Water level should be about $\frac{1}{2}$ in below the overflow outlet.

A more common failing is *continuous siphonage*. After the cistern has been flushed the siphon fails to 'break'. The cistern doesn't refill and a stream of water continues to flow down the flush-pipe. Pulling the chain for a second time breaks the siphon and permits the cistern to fill.

A combination of circumstances may cause this trouble. After years of use the lugs at the base of the bell can wear down and, at the same time, an accumulation of rust and other debris collects in the bottom of the well. As a result the gap through which the air must pass to break the siphon is reduced and, coupled with a high water pressure and an efficient ball valve that refills the cistern rapidly, continuous siphonage results.

A permanent 'cure' can usually be effected by cleaning out the well. In some cases it may be necessary to build up the lugs on the rim of the bell with an epoxy resin filler. Sometimes, too, you will need to reduce the flow of water into the cistern by partially closing a stop-cock on its supply pipe.

The intrinsic noisiness of these cisterns – the clank of the bell, the rush of water descending from high level and the hiss of incoming water direct from the main – cannot be eradicated.

Direct action flushing cisterns

Modern direct action flushing cisterns may be made of enamelled cast iron, of plastics or of ceramic material. They may be installed at either high or low level. Most have a flat base, though some well-bottomed models have been produced to simplify the replacement of old *Burlington* cisterns.

A stand-pipe rises from the base but, instead of terminating open-ended above the water line, it is bent over and opened out to form a dome, the open base of which is about

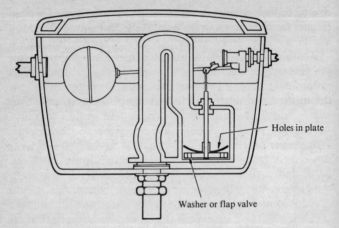

Fig. 29   A direct action flushing cistern

an inch above the bottom of the cistern. A rod, connected
to the flushing mechanism, passes through this dome. To
its lower end a circular metal or plastic plate is connected.
This plate has one or more holes in it which allows water
to flow through freely once siphonic action has started.

Immediately above the plate is a round *siphon washer* (or
flap valve) made of thin plastic. When the flushing mechanism is operated the plate is raised, water pressure holds the
plastic flap valve against the holes in the plate to close them
and water is thrown over the inverted 'U' bend into the
flush-pipe.

This falling water produces a partial vacuum and induces
a siphonic action that results in the contents of the cistern
being discharged down the flush-pipe to cleanse the pan.
Water pressure from beneath the plate pushes the flap valve
upwards to open the holes and to permit water to flow
through them.

Failure of the flap valve will result in increasing difficulty
in flushing the cistern. The holes in the plate will not be

completely closed as the plate is raised. Water will be able to pass through and two or more sharp jerks on the flushing handle will be needed before siphonage can be induced.

Renewing the flap valve is usually quite a straightforward job. Tie up the float arm of the cistern's ball valve so that no more water can flow in, then flush to empty. Unscrew the nut connecting the flush-pipe to the siphon outlet and

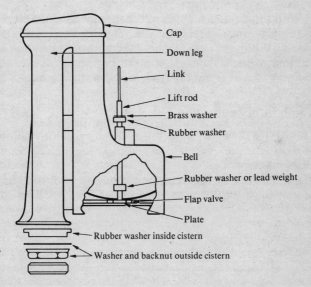

Fig. 30   Siphoning mechanism from direct action flushing cistern

pull the top of the flush-pipe away from this outlet. Next unscrew the large nut immediately below the cistern and remove it from the threaded 'tail' of the siphon. Have a bowl handy as you do this, as the pint or so of water left in the cistern will escape.

Having removed this nut you can now withdraw the siphon from the cistern. Remove the metal link that connects the operating rod to the flushing handle and withdraw the plate and rod from the base of the siphon dome. You can now remove the old flap valve and renew it.

When ordering a new flap valve tell the supplier the make of the cistern so that he can supply one of the correct size. If this isn't available ask for the biggest he has; it can easily be cut to size with a pair of scissors. It should be sufficiently large to cover the plate and touch, but not drag on, the walls of the siphon dome.

With some suites the siphon is not retained in the cistern in this way. Instead there may be retaining bolts inside the cistern. With very modern suites the flush-pipe may not be screwed to the siphon outlet at all but be thrust into the cistern past an 'O' ring that gives a watertight seal.

The lavatory pan

The straightforward 'wash down' lavatory pan is the one in most common use and the one most likely to be found in the home.

It is cleansed by the weight and momentum of the 9 litre (2 gal) flush that enters the pan via a *flushing horn* and

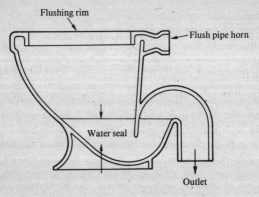

Fig. 31   Wash-down lavatory pan

passes, in two equal streams, round the flushing rim to meet at the centre of the front of the rim. Failure to cleanse the pan properly can be an embarrassing fault that can have a number of causes.

There must be *enough* flushing water. Check that the cistern fills up to the mark on its wall or, if there is no mark, to within about ½ in of the overflow pipe.

During periods of drought Water Authorities urge house-holders to reduce the volume of flush, and thus to save water, by placing a brick or a plastic bag filled with water in the cistern. When the lavatory suite is used as a urinal only the volume of the flush can safely be halved. However, when the suite is used for other purposes the full 9 litre flush may well be needed. No water will be saved if the flushing water has been reduced to such an extent that it is necessary to flush twice to cleanse and clear the pan.

Another possible cause of poor cleansing is an obstruction at the point at which the flush-pipe connects to the pan or in the flushing rim. Make sure that the flush-pipe connects squarely to the flushing horn. It used to be the practice to make this connection with an unhygienic 'rag and putty' joint. Putty was often extruded from a joint of this kind to produce an obstruction. 'Rag and putty joints' should be replaced with modern rubber cones.

Check with the fingers, or with a hand mirror, that the flushing rim is free from scale or flakes of rust. For efficient flushing the two streams of water must meet at the centre of the front of the pan. There should be no 'whirlpool effect' as the water flows out.

Check with a spirit level that the pan is set dead level on the floor. If it is not, pack under the lower side (you can use pieces of linoleum or strips of wood).

Finally, an obstruction at the outlet can prevent proper cleansing. Jointing material may be extruded from the joint between the lavatory pan and the branch drain or soil-pipe. In this event water level will rise in the pan before the contents finally flush away.

If water level in a pan that has previously flushed satisfactorily, rises almost to the flushing rim and then – very slowly – subsides, a drain blockage is the most likely cause – see pp. 134–7.

Siphonic lavatory suites

Siphonic lavatory suites depend on siphonic action – the weight of the atmosphere – to cleanse the pan. This permits the use of the neat, 'close coupled' lavatory suite in which flushing cistern and pan form one compact unit and in which no flush-pipe is visible.

The simpler kind of siphonic suite is called a *single trap* suite. The outlet of the pan is first somewhat constricted

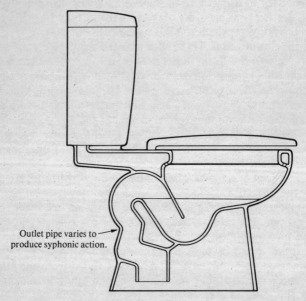

Outlet pipe varies to
produce syphonic action.

Fig. 32   Single trap siphonic lavatory suite

and then opened out before connection to the branch drain or soil-pipe.

When the cistern is flushed the constricted part of the outlet fills completely with water. This pushes air in the wider section in front of it to produce the partial vacuum on which siphonic action depends.

When flushed, the water level in a single trap siphonic suite will first rise slightly and then fall very quickly as siphonic action begins. There may be a noisy gurgle as air passes through the trap to break the siphon.

Double trap siphonic suites have a particularly effective and silent action. They are recommended where efficient unobtrusive service is a primary consideration.

A small pipe (or *pressure reducing device*) connects the

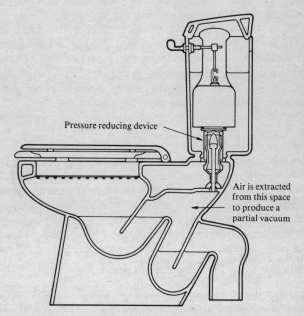

Pressure reducing device

Air is extracted from this space to produce a partial vacuum

Fig. 33 Double trap siphonic lavatory suite

air space between the two traps to the outlet of the flushing cistern. Flushing water, rushing past, sucks air out of this air-space to produce a partial vacuum – and siphonic action is produced.

When working properly, water level in the lavatory pan can be seen to fall under the pressure of the atmosphere *before* the flushing water reaches the pan. Failure to work properly is almost always due to the pressure reducing device having become blocked with debris.

Noisiness

Nobody wants a lavatory suite that is audible throughout the house. Noise can arise from the cistern refilling after use, from the descent of the flushing water or from the contents of the pan flowing into the soil-pipe.

Ways in which ball valves can be silenced to give a less obtrusive refill of the cistern are described on pp. 60–62. So far as the descent of the flushing water is concerned it should be borne in mind that low level suites are always more silent than high level ones and that double trap siphonic suites are almost silent in operation.

Noise resulting from the contents of the pan flowing into the soil-pipe is most likely to occur where, in a modern home, an upstairs lavatory is connected to a soil-pipe inside the fabric of the building. The connection between the lavatory pan and the socket of the branch soil-pipe should always be made with a mastic – not a rigid cement – joint. If this noise is a problem relief may be obtained by raising a floorboard in the lavatory compartment and running a layer of fine sand on to the ceiling of the room below.

Leakage from lavatory drain connection

Leakage from the joint between the outlet of the lavatory and the soil-pipe socket to which it is connected should never be ignored.

It is most likely to occur where an upstairs lavatory has been connected to the soil-pipe socket with a putty joint. Over the years the putty dries and shrinks. Movement of the floorboards produces cracks that allow the joint to leak when the lavatory is flushed.

To cure this, rake out all the existing jointing material with the blade of a screwdriver or similar tool. Wrap two or three turns of a waterproof building tape round the lavatory outlet and caulk down hard into the socket.

Fill the remaining space between lavatory outlet and socket with a non-setting mastic, such as *Plumbers Mait*. Complete the joint by wrapping round it a further couple of turns of building tape. This can be painted if desired.

An alternative cure, but one which involves unscrewing the lavatory pan from the floor and moving it forwards, would be to replace the existing joint with a patent plastic push-on joint such as the *Multikwik* drain connector (Phetco Ltd, Totton, Southampton).

Condensation on lavatory suites

Condensation – beads of moisture that may give the impression that the appliance is leaking – may form on both flushing cisterns and lavatory pans.

It is caused by warm, moist air coming into contact with the cold surfaces of the lavatory suite. This cools the air and, since cold air contains less moisture than warm, the surplus moisture condenses on the cold surface.

The best remedy lies in improved ventilation. In some cases it may be worthwhile fitting an electric extractor fan in the window of the bathroom or lavatory compartment.

An alternative approach is to insulate the surface of the cistern either against the warm, moist air or against the cold water that the cistern contains.

Cast iron cisterns can be coated externally with two or more coats of anti-condensation paint. Plastic cisterns have a built-in insulation that affords them considerable protection. Ceramic cisterns often give serious trouble. As a last resort consider insulating them internally. This can be done by emptying the cistern and drying it thoroughly. Then line with strips of thin expanded polystyrene sheet (sold as wall paper lining). Use an epoxy adhesive and do not refill the cistern until the adhesive has set thoroughly.

Lavatories situated in bathrooms are more prone to condensation than those provided in a separate compartment. Don't drip-dry clothes over the bath and do run a couple of inches of cold water into the bath before turning on the hot tap – this will all help.

6

Hard, Soft and
Corrosive Water Supplies

The oceans that cover two-thirds of the surface of the earth
are ultimately the world's great reservoirs of water supply.
Water evaporates from the surface of the ocean and is blown
over the land. Rising, to pass over hills and mountains, the
warm moisture-laden air cools. It can no longer retain its
content of water vapour, which condenses and falls as rain.

Much of the rain flows, via streams and rivers, back
to the ocean; some flows into lakes and reservoirs; some
penetrates the surface strata to collect in vast natural underground reservoirs.

Water has the property of dissolving virtually any solid
or gas with which it comes into contact. When it leaves the
rain-cloud it is naturally distilled. On its brief journey to
the earth's surface it dissolves measurable amounts of
carbon dioxide and, over industrial towns, sulphur dioxide
and other by-products of industrialisation.

On reaching the earth the process continues. Water
flowing over moorland takes into solution chemicals
produced from dead vegetation and is likely to have acid
characteristics. Water permeating through the surface rocks
into natural underground reservoirs will dissolve some part
of these rocks as it does so. Such water is likely to contain,

in solution, the bicarbonates, sulphates and chlorides of calcium and magnesium. These are the chemicals that produce *hardness*.

Hardness

Hardness is usually measured in terms of the equivalent of calcium carbonate in the water in parts per million (expressed as p.p.m.). Occasionally *degrees of hardness* are referred to. To convert degrees of hardness to p.p.m. of calcium carbonate, multiply by fourteen.

Water in which is dissolved the equivalent of more than 100 p.p.m. of calcium carbonate is described as *moderately hard*; if it has 200 p.p.m. or more – hard. 65% of Britain's homes are supplied with either moderately hard or hard water and are found mainly in central, southern and eastern Britain. The Scottish Highlands and Wales generally have soft water supplies though, even there, there are some hard water sources.

A chemical analysis of a local water supply will yield three measurements of hardness: *temporary* hardness, *permanent* hardness and – the sum of the two – *total* hardness.

Temporary hardness means hardness that can be removed by boiling. It is caused by the dissolved bicarbonates of calcium and magnesium. Permanent hardness is caused by dissolved sulphates and chlorides of these elements.

The ill-effects of temporary hardness have been discussed in Chapter 3. Insoluble calcium and magnesium carbonate are precipitated out as *fur* or boiler scale at temperatures of 70°C (160°F) and above. To check whether or not your local water supply has temporary hardness, look inside your kettle. If its internal surface and the electric element have a coating of creamy white *fur* you can be sure that your water supply has a degree of temporary hardness.

The formation of boiler scale is not the only disadvantage of a hard water supply. Hard water will not dissolve soap readily. An insoluble 'curd' is formed that matts woollens, makes hair washing a misery, produces a dirty tide mark in baths and basins and – of course – wastes soap.

Although only temporary hardness causes hard water scale the chemicals responsible for both temporary and permanent hardness are left behind when hard water evaporates. These can clog ball valves and leave unsightly marks round the nozzles of taps and on bath and basin surfaces.

Water softening

Water can be reduced to zero hardness by the use of a *base exchange* or *ion exchange* water softener.

The principle of ion exchange water softening was discovered by observing the softening effect that occurred when hard water flowed through natural zeolite sands, and it was these sands which were used in early water softeners.

Nowadays synthetic resins are used but the principle remains the same. At the risk of over simplifying a complex chemical reaction it may be said that as hard water passes through an ion exchange water softener, the chemicals that produce hardness *exchange bases* with chemicals in the resin. For instance, dissolved calcium bicarbonate is changed into dissolved sodium bicarbonate which does not cause hardness.

After a period of time the softening qualities of the resin are exhausted. They can however be renewed by passing through the softener a strong solution of sodium chloride (common salt).

The cycle of regeneration must be undertaken at regular intervals – perhaps once a week in a household with average water consumption. The resin is 'back-washed' to loosen up the resin bed and remove any debris. The back-wash water runs to waste. A strong brine solution is then passed through the resin. A final rinse, also run to waste, washes surplus brine from the resin bed before the softener is brought back into normal use.

This regular chore was a disadvantage of early softeners. However modern appliances have a substantial reservoir containing sufficient salt for anything between fifteen and thirty regenerations. The cycle of regeneration takes place automatically at preset intervals, regulated by a time clock.

As well as the large – and expensive – mains water softeners there are also smaller portable softeners which operate on the same principle. These have a hose connector for attachment to a kitchen or bathroom tap and can be used for softening relatively small volumes of water. A large screw cap at the top of the softener gives access for salt when the resin needs to be regenerated.

Small quantities of water can also be softened by dissolving certain chemicals in them. Most of these are based on sodium carbonate (washing soda). However, these produce a precipitate of hard water chemicals and can sometimes irritate sensitive skins.

Sodium hexametaphosphate (sold commercially as *Calgon*) is a water softening chemical that works rather differently. It does not produce a precipitate but combines with the chemicals causing hardness and neutralises them. *Calgon* is more expensive than soda-based softening chemicals but can safely be used whenever soft water is required.

Hard water has its merits

Those who live in an area with a hard water supply may console themselves with the knowledge that hard water is not without its advantages nor soft water without its drawbacks.

Iron pipes – very subject to corrosion in soft water areas – will acquire an internal coating of *egg-shell* scale where the water supply is hard that gives a considerable measure of protection.

Soft water dissolves metals much more readily than hard. This is particularly worrying where lead pipes are in use as its cumulative effect can produce serious ill-health. [It has even been suggested that malaise resulting from the use of lead water distribution pipes contributed to the fall of the Roman Empire!] Water drawn from lead service pipes in soft water areas has been found to have dangerously large amounts of this metal dissolved in it.

Possibly connected with this property of dissolving metals is the fact that there is strong evidence that diseases of the heart and circulatory system are more prevalent in soft than in hard water areas.

Then too, many people prefer the taste of hard water – particularly when it is used for tea or coffee making. For these reasons, in my opinion, it is best to install a mains water softener into the rising main *after* the branch that takes water to the cold tap over the kitchen sink.

This means that while soft water will be available in the bathroom and from the kitchen hot tap, hard water will continue to be used for drinking and food preparation.

If you live in an area with a naturally soft water supply it is wise, particularly if you have a lead rising main, to allow a few pints of water to run to waste before you fill the kettle first thing in the morning. It is overnight, when water has been left standing in the rising main, that it will have taken in lead into solution.

Corrosion

Corrosion – the chemical effect of water plus oxygen on metals – is closely associated with the solvent properties of water. [The atmosphere in north-western Europe always contains water vapour and, under normal circumstances, water always contains dissolved oxygen.] Virtually all metals are affected to some extent by corrosion but, in most cases, this has little or no effect upon their use. Lead and copper pipes, for instance, if exposed to the atmosphere, quickly acquire a thin coating of lead or copper oxide which dulls their appearance. However this coating then protects the metal from further attack.

Iron and steel are affected differently from copper and lead. Iron oxide (rust) forms on the surface. This expands and flakes away leaving the metal beneath exposed to further attack. It is therefore iron and steel plumbing equipment that is principally in need of protection from corrosion.

External iron-work – gutters, down-pipes, soil-pipes and so on – are protected by painting. The paint prevents the metal from coming into contact with the atmosphere. Its protective effect is far more important than its decorative one. It is therefore important to ensure that this protection extends to those parts of external iron work that are out of sight – the backs of rain water down-pipes and the insides of rain water gutters for instance. Routine external painting should always include the coating of the out-of-sight internal surfaces of rain water gutters with a protective bituminous paint.

The cold water storage cistern and electrolytic corrosion

The traditional mild steel cold water storage cistern which, despite the increasing use of non-corrodible plastic cisterns, is still to be found in many thousands of British homes, is customarily protected from corrosion by galvanising. This is an electro-chemical process which coats the surface of the cistern with a thin layer of zinc.

Under normal circumstances zinc forms a coating of zinc oxide in the presence of air and water so protecting it from further corrosion. However, the circumstances to which the zinc internal coating of a cold water storage cistern are exposed are not necessarily 'normal'.

A simple electric cell consists of a rod of copper and a rod of zinc in electrical contact with each other immersed in an acid solution (referred to as the *electrolyte*). Electric current passes from one rod to the other, bubbles of hydrogen gas are produced in the electrolyte, and the zinc rod gradually dissolves away.

These conditions may be reproduced in a galvanised cold

water storage cistern where the rising main and distribution pipes are of copper and the water supply is slightly acid. The water in the cistern acts as an electrolyte. Electric current passes between the zinc coating of the cistern and the copper pipes attached to it. The zinc dissolves to leave the mild steel underneath unprotected.

This process, known as electrolytic corrosion, has become common in post-war years as a result of the almost universal adoption of copper tubing for water supply pipes.

Protecting the cold water storage cistern
Galvanised steel cold water storage cisterns can be protected from the effects of elctrolytic corrosion by painting them internally to prevent the water that they contain from coming into contact with the vulnerable metal surfaces.

New cisterns should be treated *after* the holes have been cut for the pipe connections but *before* these connections are made. Rub down the internal walls of the cistern with abrasive paper to form a *key* for the paint. Carefully remove every trace of metal dust or shaving.

Apply two coats of a *tasteless and odourless* bitumastic paint. Cover the entire interior of the cistern, paying particular attention to the bare metal of the holes cut for the pipes. This treatment will give several years' protection and can be repeated if necessary.

An existing cistern, already affected by corrosion, can be reconditioned and protected in the same way. Cut off the water supply to the cistern, drain and dry it. As the water distribution pipes will be taken from a point about two inches from the base of the cistern you'll have to bale and mop out the final gallon or so of water.

While it is not absolutely essential to disconnect the supply, overflow and distribution pipes it is best to do this if possible.

Remove every trace of existing rust. This can be done by wire brushing (wear goggles to protect the eyes) or chemically. In the past I have had some doubts about the value of chemical rust removers but I have found the biological removers currently on the market to be extremely effective.

Removing the rust may leave pit-marks or even holes in the walls of the cistern. Fill these with an epoxy resin filler. When set, rub down to a smooth surface before applying two coats of bitumastic paint.

Anodic protection

An alternative method of protection, that can be used where a cistern is not yet showing signs of corrosion, is the provision of a *sacrificial anode* in the storage cistern.

All metals have a fixed electric potential. When electrolytic action takes place it is the metal with the higher potential that dissolves away. The zinc coating of a galvanised storage cistern has a higher potential than copper, but magnesium has an electric potential that is higher still.

A sacrificial anode therefore consists of a lump of magnesium. One well-known make is sold complete with a length of copper wire attached to it and, at the other end of the wire, a clamp for connection to the rim of the cistern.

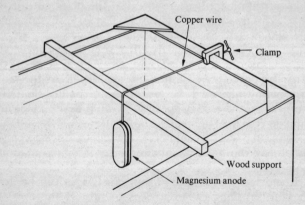

Fig. 34   A magnesium anode in a cold water storage cistern

To fit, rub down the rim of the cistern at the point where the clamp is to be fitted to ensure contact with the bare metal. Then lay a batten of wood across the cistern and hang the copper wire over it so that the magnesium anode is suspended deep in the water of the cistern.

Now, when electrolytic action takes place, the magnesium anode will dissolve away and the zinc lining of the cistern will be protected.

Anodes of a rather different design are also produced to protect from corrosion the galvanised steel hot water tanks that were often used in pre-war – and immediately post-war – *cylinder storage* hot water systems. These hot water tanks cannot be protected by internal painting. Anodes designed for their protection consist of a lump of magnesium on the end of a metal rod.

To fit one, the hot water system must be drained and the hand-hole cover of the tank unbolted and removed. A hole is then drilled through the centre of the hand-hole cover and the rod fitted into this hole so that it projects with its anode into the middle of the tank.

When fitting a new galvanised steel hot water tank it is essential to make sure that every trace of metal dust or shaving is removed before the tank is connected up. Any trace left in the tank will unfailingly become a focus of corrosion.

It should now be obvious that a galvanised steel plumbing system should never be extended with the use of copper tubing. Electrolytic corrosion could result in the early failure of the galvanised steel.

Where such a system is to be extended it is best to use stainless steel tubing. This is almost as easy to handle as copper tubing and – whether used with copper or galvanised steel – does not involve any risk of electrolytic action.

Dezincification

In some parts of the country the corrosive nature of the

water supply produces a phenomenon known as *dezincification*. This is an electro-chemical action that affects brass fittings.

Brass – commonly used for stop-cocks, ball valves, drain-cocks and compression joints and fittings – is an alloy of copper and zinc. When dezincification takes place the zinc is extracted from the brass leaving the fitting unchanged in appearance but without any structural strength.

In areas where dezincification occurs it is wise to use plumbing fittings made of *gunmetal* (an alloy of copper and tin). These are readily available. Any local builders merchant will tell you whether, in your particular area, good class installers customarily use brass or gunmetal fittings.

Boiler corrosion

Rusty water running from a hot water tap – particularly noticeable when a large volume of water has been run off for a bath – indicates serious corrosion somewhere in the hot water system.

First of all check the condition of the cold water storage cistern and the interior of the galvanised steel hot water storage tank if you have one. If these are both rust free the trouble must exist in the boiler itself. This demands immediate attention, as a rusting boiler will eventually become a leaking boiler.

The introduction of Micromet crystals (see p. 20) into the cold water storage cistern will sometimes help. The most certain remedy however is to convert the hot water system from direct to indirect operation.

As explained in Chapter 3 the water in the primary circuit of an indirect system is never drawn off. It circulates continuously between the boiler and the storage cylinder. When this water is first heated the air dissolved in it is driven off. Thereafter there will never be sufficient air present in the primary circuit to produce boiler corrosion.

Internal corrosion in central heating systems

Although, the boiler of an indirect hot water system is never likely to corrode, a kind of electrolytic corrosion can take place as a result of an electro-chemical reaction between the pressed steel radiators and the copper circulating pipes of a central heating system – despite the fact that this central heating circulation will be connected to the primary circuit.

Very small amounts of air will be present in solution. Some will enter by way of minute leaks at joints which are too small to permit water to escape. Some will dissolve into the surface of the water in the feed and expansion tank. This tiny amount of air will be sufficient to induce the kind of corrosion referred to.

The first indication of internal corrosion is usually that one or more radiators fail to heat up and need frequent venting or 'bleeding' to restore them to service.

To vent a radiator, insert a key into the air vent at a top corner of the radiator – usually the corner opposite to the radiator inlet. Hold a cloth under the vent to catch any escaping water and turn the key. At first gas will escape. When it is followed by water, turn off the vent.

If a radiator needs frequent venting, apply a lighted taper to the escaping gas. Air will not burn but if it is hydrogen – one of the products of corrosion – it will burn with a blue flame.

The other product of corrosion is black oxide sludge or *magnetite*. This clogs the bottom of radiators and obstructs circulating pipes. It is drawn to the magnetic field of the electric circulating pump and its abrasive characteristics are a frequent cause of early pump failure.

Both hydrogen gas and magnetite are produced from the walls of the pressed steel radiators. It won't be long before these begin to leak. This internal corrosion can be prevented by introducing a suitable corrosion inhibitor into the feed and expansion tank of the system.

It is essential that an existing system should be thoroughly

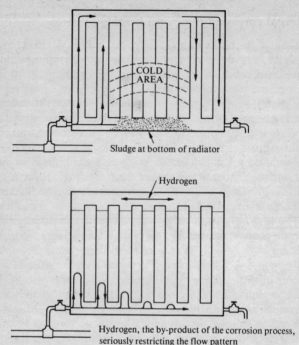

Fig. 35   The effects of internal corrosion on radiators

flushed through to remove existing sludge before the inhibitor is introduced. One way of doing this is to connect a hose to the drain-cock of the boiler and to take the other end to an outside drain gully. Turn off the boiler and open the drain-cock. Do *not* tie up the ball valve of the feed and expansion tank or otherwise prevent water flowing into this tank.

Go to each radiator in turn, beginning with that furthest from the boiler and tap the surface with a padded mallet to loosen any adhering sludge. Do not close the drain-cock until the water flowing from it has appeared clear for at least twenty minutes. This flushing process can be made easier by the use of proprietary detergent solutions.

Next, tie up the float arm of the ball valve supplying the feed and expansion tank. Draw off two buckets full of water from the boiler drain-cock and pour the corrosion inhibitor into the feed and expansion tank.

Draw off a further bucket full of water from the drain-cock to ensure that the inhibitor is drawn into the system. Untie the float arm of the ball valve and permit the feed and expansion tank to fill to its normal level. Then run the central heating system for at least an hour to allow the inhibitor to circulate throughout the system.

A 'fringe benefit' of corrosion inhibitors is the way in which, by altering the resonance range of the central heating system, they will often reduce or eliminate noise from the operation of the system.

Self-priming indirect systems

A snag about self-priming indirect systems (see Chapter 3) is the difficulty of protecting them from internal corrosion. This *can* be done by draining the system down, removing the hand-valve inlet from one of the radiators, and introducing the corrosion inhibitor directly into that radiator using a funnel and a length of rubber pipe.

Manufacturers of chemical corrosion inhibitors do not recommend this course of action though. They feel that self-priming cylinders give an insufficiently positive separation of the primary water and the domestic hot water, and there could be a risk of contaminating the domestic hot water with the chemical inhibitor.

7

Frost and the Plumbing System

The domestic hot and cold water systems are under continuous assault from corrosion and hard water scale. But
because frost only presents a danger for a few weeks in
every year in the British Isles, frost precautions are often
taken far less seriously here than in countries where weeks
– or months – of sub-zero temperatures are the norm. Yet
the danger from frost is real enough. This was amply demon-
strated during the winters of 1963/1964 and of 1978/1979
when the plumbing systems of thousands of British homes
froze and burst. Even during our 'normal' mild winters
many householders are caught out by a sudden and un-
expected cold snap.

Water's unique physical characteristics

Water, in common with all other liquids, expands when
heated. The satisfactory functioning of cylinder storage hot
water systems (discussed in Chapter 3) depends on this
property. The expansion is fairly modest in dimensions.
Water raised from 4.4°C (40°F) to 100°C (212°F) increases
in volume by one twenty-third.

At that temperature, water changes state to become a gas

(steam) and expansion is no longer modest. Steam has a volume 1,689 times that of the water from which it was produced. Thus, one cubic foot of water produces 1,689 cubic feet of steam!

Water also has the unique quality of expanding when cooled below 4.4°C (40°F). Gradual expansion takes place between 4.4°C and 0°C (32°F). At 0°C water again changes state to become a solid and increases in volume by approximately one-tenth. Further expansion takes place if the temperature is lowered further.

These characteristics are of great importance in understanding the effects that frost may have upon the plumbing system. The one-tenth expansion of water on freezing accounts for the burst water supply and distribution pipes that follow any severe period of frost. The 1,689% increase in volume that takes place when water turns to steam accounts for the devastating effects of the boiler explosions that can occur.

Frost precautions – design considerations

The most effective defences against frost are those that are built into the house when it is designed and constructed. Some of these have been referred to briefly in earlier chapters.

The service pipe from the water main to the house boundary should be protected by a covering of at least 80 cm (2 ft 6 in) of soil. Where it passes through a hollow underfloor space it should be threaded through a 15 cm (6 in) drain pipe filled with insulating material.

The rising main and distribution pipes should, as far as possible, be positioned against internal, rather than external, walls. Pipe runs in the roof space should be short and kept well away from the eaves.

The cold water storage cistern should be positioned against a flue in regular use. The boiler, hot water cylinder and cold water storage cistern should preferably form a vertical column, one above the other, so that warm air rising from the boiler and cylinder can protect the vulnerable cold water storage cistern. These measures will positively *add* warmth to the plumbing system whereas even the most thorough lagging will do no more than reduce the rate of heat loss.

Lagging

This is not, of course, to suggest that lagging is of little or no importance. In an occupied house, with the plumbing fittings in regular use, reduction of the rate of heat loss will be all that is needed to protect the water supply and distribution pipes.

Inorganic lagging materials, which will not harbour household pests, should be used. Fibreglass pipe-wrap is suitable for out-of-sight pipe runs but, where the pipework is visible, expanded polystyrene or foam plastic pipe lagging units are usually preferred.

Where pipes are situated against external walls it is most important that the lagging should extend behind the pipe. It will be to the cold external wall – not to the warm air of the room – that the pipe will lose warmth.

All pipework in the roof space should be thoroughly lagged. This is of particular importance where the householder has insulated the attic, to reduce heat losses from the rooms below. As a result the bedrooms will be warmer – but the roof space will be that much colder.

Make sure that the lagging covers the bodies of any gate valves or stop-cocks, leaving only the handles exposed. Don't forget to lag the vent pipe of the hot water system. Remember that water will stand in this pipe to the same level as that in the cold water storage cistern.

The cold water storage cistern should have a tight – but not air-tight – fitting cover as a protection against both frost and contamination. Galvanised steel and asbestos cement cold water storage cisterns can be lagged either with fibreglass tank-wrap or with expanded polystyrene tank

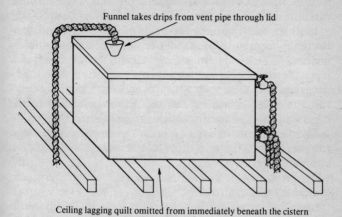

Funnel takes drips from vent pipe through lid

Ceiling lagging quilt omitted from immediately beneath the cistern

Fig. 36 A lagged galvanised steel cold water storage cistern

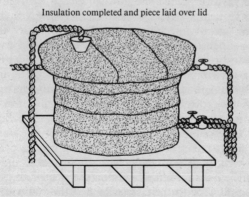

Insulation completed and piece laid over lid

Fig. 37 A lagged round polythene storage cistern

lagging units. Plastic cisterns have a certain built-in protection against frost but lagging remains a worthwhile precaution.

Lag the walls, but not the base, of the cistern and omit the fibreglass quilt or loose fill from the ceiling immediately beneath it. This will permit some warmth to escape upwards from the room below. It is advantageous to continue any fibreglass tank lagging downwards to form a kind of funnel through which warm air can rise from the ceiling to the cistern without being dispersed through the roof space.

The vulnerable overflow pipe of the storage cistern was referred to in Chapter 2. It should be bent over within the cistern to extend an inch or so below the surface of the water. A trap will thus be formed that will prevent cold draughts from entering.

Outside lavatories are very exposed to attack by frost. One way of protecting them from damage is to fit a stopcock into the cold water supply pipe just above ground level. Turn it off when a frost is forecast and flush the cistern empty. this will not prevent the pipe from freezing but, with the upper end of the pipe open, ice will be able to expand longitudinally up the pipe instead of laterally against its walls which should then escape damage.

Adding heat by electro-thermal lagging gives the most positive means of protection. Purpose-made electrical heating tape is wound round the supply pipe. This is then plugged into a convenient power socket and switched on when a frost risk is imminent.

Most outside lavatories do not, of course, have a convenient power socket. Electro-thermal protection can be improvised using the lavatory's electric light socket. Fit a light socket plug to one end of a length of two-core cable and a light socket with a 100 watt bulb to the other end. Remove the bulb from the lavatory light socket and replace it with the light socket plug. Loop the electric flex over the cold supply pipe to the cistern so that the 100 watt bulb is suspended a foot or so below the ball valve inlet (*outside* the cistern of course!). Draught-proof the lavatory compartment thoroughly. Switching on the light bulb will then generate sufficient heat to protect the cistern and water supply pipe from several degrees of frost.

Frost and the drainage system

The Building Regulations, produced in the aftermath of the severe frosts of 1963/1964, included the requirement that all soil and waste pipes taking drainage from the upper floors of buildings should be contained within the fabric of the building. This regulation hastened the now almost universal adoption of the *single stack drainage system* (see Chapter 8) and resulted in a considerable improvement in the external appearance of British homes built after the Regulations came into effect.

As far as protection is concerned however it was something of a panic measure and has recently been amended to permit external soil and waste pipes where a building is no more than three storeys high.

Waste water is always discharged at a temperature several degrees – usually many degrees – above freezing point. It will not freeze under even the severest conditions that are experienced in the British Isles during its brief journey to the underground drain.

The risk of a waste pipe becoming blocked with ice arises, not from any failing of the drainage system, but from inadequate maintenance of hot or cold water taps. If, because of a worn washer or because it is believed that, 'running water won't freeze', a tap is allowed to drip throughout a night of severe frost, a layer of ice can build up round the internal walls of a waste pipe to produce a completely blocked pipe by morning. Attention to tap washers should therefore be a routine preparation for the winter and the householder should make sure that all taps are securely turned off at night.

If frost does strike

If, despite all your precautions, you should get a freeze-up in your plumbing system the secret of success is to deal with it promptly. At first there will be only a small ice-plug, easily melted, in a supply or distribution pipe. If it is neglected it will quickly spread throughout the pipe and thawing out will become a much more difficult task.

Try to establish the position of the ice plug. You will know which pipe is affected by which appliances fail to work. If, for instance, water flows freely into the main cold water storage cistern but the lavatory cistern fails to refill after flushing then the blockage must be in the distribution pipe between the main cistern and the flushing cistern. If the cold tap over the sink is working normally but the main storage cistern is not filling the blockage must be in the rising main between the cold tap and the storage cistern.

An inspection will reveal the most likely position. It could be at a point where lagging has come away from the pipe, at a point where the pipe is fixed closely to an external wall or it could be in the roof space – possibly near the eaves.

Strip off the lagging and apply cloths dipped in hot water and then wrung out. Hot water bottles can also be useful in applying heat to a frozen pipe and an electric hair dryer can be used to direct a current of warm air towards an otherwise inaccessible blockage. Do not risk starting a fire by using a blow torch – particularly on pipes fixed to the dry timbers of the roof space.

Dealing with a burst pipe

Pipes burst as a result of the 10% expansion that takes place when water turns to ice. The burst occurs as the water in the pipe freezes but it will remain undiscovered until water starts to flow again. (Hence the widespread belief that it's the thaw that bursts pipes.)

The first sign of a burst is likely to be water dripping through a ceiling.

Before making any move to discover the source of the trouble take immediate steps to limit the damage that it could cause by turning off the main stop-cock and opening up all kitchen and bathroom taps.

Only after you have taken these steps should you look for the burst pipe.

If you have a modern copper plumbing system it is likely that there will be no actual 'burst'. Expansion of the ice will have forced open a compression or soldered capillary joint somewhere along a run of pipe. The way in which these joints are made – and may be remade if necessary – is described in Chapter 11.

Lead pipe will probably have split. The traditional method of repairing a split lead pipe is to cut out the affected length and to insert a new length of pipe, connecting the ends to the sound part of the old pipe with *wiped soldered joints.*

The technique of making a wiped soldered joint is described in Chapter 11 but it must be stressed that this demands considerable practice. Faced with a burst lead pipe most householders would be wise either to seek professional help or to make a temporary repair with an epoxy resin repair kit that can be obtained from d-i-y and motor accessory shops.

The edges of the split should be knocked together with a hammer and the area of the pipe cleaned thoroughly with a medium abrasive paper. Mix up the resin and hardener according to the maker's instructions and spread generously all round the pipe in the area of the split. Bind with a nylon bandage while it is still plastic and cover this bandage with a further layer of filler. Leave until set and then re-establish the water supply. When set the repair can be rubbed down with fine abrasive paper to give a neat finish.

Some people imagine that because a lead pipe has been frozen on one or more occasions without obvious damage, it will never burst. This is far from being so. Lead is a soft metal with little elasticity. When ice forms in a lead pipe it will expand, increasing the pipe's diameter – but the walls of the pipe will, as a result, be a little thinner than they were before. When the ice melts the pipe will not resume its former size. This process may be repeated two or three times before a burst actually occurs but, with each freeze-up, the pipe wall has become thinner and so more likely to split.

Leaving the house during the winter months

It was pointed out earlier in this chapter that lagging cannot *add* warmth to the plumbing system, it can only slow down the rate of heat loss. Provided that the house is occupied and the plumbing fittings are in normal use, this is all that is required. Water entering the domestic cold water system from the Authority's main is always a few degrees above freezing point. Before its temperature can fall to 0°C it will be drawn off and replaced by more, marginally warmer, water from the main.

The space heating of the house will further reduce the rate of heat loss, and may, in fact, raise the temperature of water flowing through the rising main and distribution pipes.

If the house is left empty and unheated during an icy spell, though, conditions will be very different. Water will stagnate in the rising main, the cold water storage cistern and the hot and cold water distribution pipes. The fabric of the house will lose its warmth and – however thoroughly the plumbing system may be lagged – a freeze-up will inevitably follow.

This fact has increased in importance in recent years. Formerly most people rarely left home during the winter except, perhaps, for a couple of days over Christmas – when the weather is usually mild. Winter holidays in the sun are now extremely popular. The householder who takes one must make sure that his plumbing system is safe before he leaves home if he is not to risk returning to a flooded house with the plumbing system in a thoroughly dangerous condition.

These precautions depend on the nature of your plumbing system. If you have a simple *direct cylinder* hot water system (see Chapter 3) the whole system should be drained. Let out the boiler fire or switch off the immersion heater. Turn off the main stop-cock and open all the taps. Because the hot water distribution pipes are taken from *above* the hot water storage cylinder, this cylinder will remain full of water after the taps have ceased to run. To drain it connect one end of a length of hose to the drain-cock beside the boiler or – if the cylinder is heated by immersion heater only – to the drain-cock at the base of the cylinder. Take the other end of the hose to an outside gully and open the drain-cock.

Having completed this operation write out large notices to remind yourself on your return that the system has been drained. *On no account must the boiler fire be lit or the immersion heater switched on until it has been refilled with water*.

If there is a drain-cock on the rising main immediately above the main stop-cock this too should be opened and the rising main drained.

Finally, just before leaving the house, flush the lavatory cistern to empty it and throw a handful of salt into the water in the pan. Salt water freezes at a lower temperature than fresh and produces a mushy ice that is unlikely to cause damage.

On no account should the primary circuit or the radiator circuit of an indirect hot water/central heating system be drained. Water drying from interior surfaces of radiators and of the circulating pump could result in serious corrosion.

If you have gas-fired central heating with automatic controls the best solution is to leave the system switched on under the control of a *frost-stat*. This is a thermostatically controlled device that will bring the system into operation when the temperature drops to a pre-determined level.

Having done this it might be wise to turn off the main stop-cock and drain the system *from the taps only*. Partially remove the trap-door of the roof space so that some warm air can reach the feed and expansion tank of the primary circuit.

If you have a central heating system that does not lend itself to automatic control of this kind (one fired by solid fuel for instance) you would be wise to introduce an anti-freeze solution into the primary circuit via the feed and expansion tank. *Automobile anti-freeze should not be used.* Manufacturers of corrosion inhibitors for central heating systems usually include suitable anti-freeze solutions among their range of products.

The anti-freeze should be added in the same way as was described in Chapter 6 for the introduction of corrosion inhibitor. Do not introduce it at the last moment before leaving home. It must have an opportunity to circulate throughout the system.

The domestic hot water circuit of a system treated in this way will not, of course, be protected from frost. This must be drained from the taps and from the drain-cock at the base of the hot water cylinder.

Boiler explosions

Boiler explosions fortunately very rarely occur and many people worry quite unnecessarily about them. They are most likely to happen where a family return from a winter holiday without having taken the precautions that I have suggested.

Supposing that, while the house was empty, there were several days and nights of extreme cold. Ice plugs may well have formed in the upper part of the hot water system's vent pipe and in the cold supply pipe from the cold water storage cistern to the cylinder. It is even possible – particularly if the boiler is situated against an external wall – that ice plugs may have formed in the flow and return pipes between the boiler and the cylinder.

What would happen if, under these circumstances, the boiler fire was lit?

Any ice in the boiler itself would melt. The resulting water would heat up, quickly reaching boiling point. It would however be unable to expand and, since it would be under pressure in a totally enclosed vessel, it would be unable to boil. Pressure would increase within the boiler until a seam or a joint gave way. Instantly, pressure would be released and the superheated water would change into steam occupying, you will recall, 1,689 times the space occupied by the water from which it was derived.

Hardly surprisingly, the boiler would explode like a bomb, with equally catastrophic results.

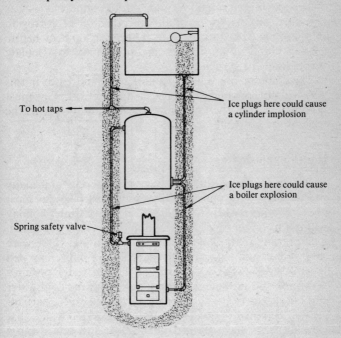

Fig. 38   A cylinder hot water system forms an open-ended U tube

A cylinder hot water system forms a kind of giant U tube with the cold water storage cistern and the vent pipe of the hot water cylinder as its open ends. Provided just one of these ends remains open there can be no dangerous build-up of pressure.

As a final safeguard most hot water systems are provided with a spring safety valve. This should be situated as close as possible to the boiler. It is usually to be found on the flow pipe though the return is equally appropriate. This will open in an emergency to release excessive pressure.

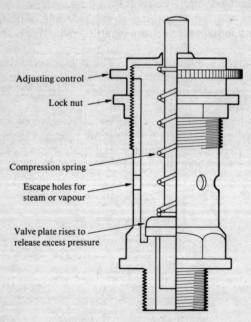

Fig. 39   A spring safety valve

Cylinder implosion or collapse

Cylinder collapse is another phenomenon that may occur during a period of severe frost. It is far more common than boiler explosion but is fortunately far less devastating.

It is likely to occur when – perhaps to save fuel or perhaps as a mistaken precaution against boiler explosion – a householder lets out his boiler fire at night. Ice plugs may form in the upper part of the vent pipe and in the cold water supply pipe to the cylinder.

Meanwhile the water in the cylinder cools and contracts, producing a partial vacuum. Copper cylinders have very little resistance to *external* pressure and the cylinder will then collapse like a paper bag under the weight of the atmosphere. Actual collapse often occurs as the householder draws water from his hot tap first thing in the morning. The further reduction in internal pressure resulting from this is the final straw!

Sometimes – but not always – the cylinder is undamaged and will resume its former shape when the ice plugs have thawed and water can once more flow down the cold water supply pipe.

Cylinder collapse can be prevented by ensuring that the vent pipe and the cold water supply pipe are thoroughly lagged – particularly within the roof space – and by keeping the boiler fire alight all night.

8

Above-ground Drainage

Providing your bathroom and kitchen appliances with a constant supply of hot and cold water is only half the role of plumbing in the home. Getting the waste water safely away again through an efficient and self-cleansing drainage system is perhaps an even more important aspect of domestic plumbing.

Traps

All plumbing appliances discharge their waste into a branch soil or waste pipe through a trap designed to hold sufficient water to prevent drain smells coming back into the room in which the appliance is installed.

The trap of a lavatory pan is an integral part of the pan, built in at the time of manufacture. All other appliances have a separate trap, made of lead, brass, copper or plastic material which is connected to the waste outlet by means of a large retaining nut.

The simplest kind of trap consists of a metal U bend. This kind is suitable if out-of-sight but where appearance is a consideration, a *bottle trap* made of chromium plated brass, or plastic, is preferred.

Traps with a vertical outlet are described as *S traps*; those with a more or less horizontal outlet as *P traps*. All traps must have a smooth internal surface to enable waste to pass through them freely.

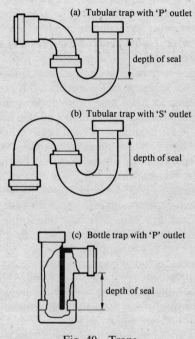

(a) Tubular trap with 'P' outlet

depth of seal

(b) Tubular trap with 'S' outlet

depth of seal

(c) Bottle trap with 'P' outlet

depth of seal

Fig. 40   Traps

A trap's efficiency at keeping back drain smells depends to some extent upon its *depth of seal*. This is the vertical distance between the overflow outlet of the trap and the upper part of the bend at its base (Figure 40). *Deep seal* traps have a 75 mm (3 in) seal. *Shallow seal* traps have a seal of 50 mm (2 in) or 38 mm (1½ in).

The two-pipe drainage system

Upstairs lavatories normally have a P trap outlet connected to the socket of a branch soil pipe by means of a flexible or mastic joint. The branch soil pipe discharges, in turn, into the main soil and vent pipe of the drain. This pipe connects at its base to the underground drain and is continued, open-ended, to above eaves level to serve as a high level ventilator. A plastic or wire 'balloon' should be fitted into its open upper end to prevent birds from nesting in it.

Ground-floor lavatories are sometimes connected to the lower part of the soil-pipe in the same way. It is more usual though for them to have an S trap outlet which may be connected to the protruding socket of a branch underground drain with either a flexible or a rigid cement joint.

The destination of waste pipes from sinks, baths, basins and bidets depends both on the location of the property and the date it was built.

Nearly all houses built before the mid-1950s were provided with a two-pipe drainage system. This system, evolved by Victorian sanitary engineers in a determined effort to keep 'drain air' out of the home, made a firm distinction between 'waste' and 'soil' appliances.

Soil appliances included lavatories, urinals and slop-sinks though, of course, only lavatories were required in domestic property. The outlets of these appliances were connected to the underground drain either directly or via a soil and vent pipe with an untrapped outlet.

Any room containing a soil appliance had to be ventilated directly to the external air or provided with means of mechanical ventilation capable of producing three air changes every hour. Furthermore an *intervening ventilated space* had to be provided between the room containing the soil appliance and any living room, bedroom or kitchen. In most cases this requirement was met simply by arranging for the lavatory compartment to be entered from a hallway or landing.

Waste appliances – sinks, baths, basins and bidets – could be installed in, or in direct contact with, habitable rooms. They were not therefore permitted to discharge directly into the drain. Their waste outlets had to discharge in the open air over a yard gully, the outlet of which was, of course, connected to the underground drain.

In a further determined effort to keep 'drain air' out of the house the bye-laws of some local authorities even prohibited the discharge of wastes directly over gullies. They had to empty into a glazed channel leading to the gully, but at least 18 in away from it.

The two-pipe drainage system presented no great difficulty so far as downstairs appliances were concerned and, when it was first evolved, the plumbing systems of *most* British homes consisted simply of an external lavatory and a shallow stoneware sink in the kitchen.

An increasing demand for homes with an upstairs bathroom meant that some means had to be devised to dispose of first floor bath and basin wastes. The most common solution – which produced results far worse than those that it was intended to cure – was to discharge these wastes over a rain-water hopper head fixed against an external wall just below the level of the bathroom floor. A length of rain-water pipe connected to the hopper head outlet took the waste to discharge over a yard gully.

Hopper heads are thoroughly insanitary appliances. They are not self-cleansing. Soapy water drying and decomposing on their internal surfaces produces unpleasant smells in the immediate vicinity of bathroom and bedroom windows. Draughts blowing up the length of rain-water pipe discharge smells from the yard gully into the same area.

Recognising this, some progressive local authorities – notably the former London County Council – forbade their use. In the London area branch wastes were required to discharge into a main waste pipe. The lower end of this discharged over a gully but the upper end had to be taken

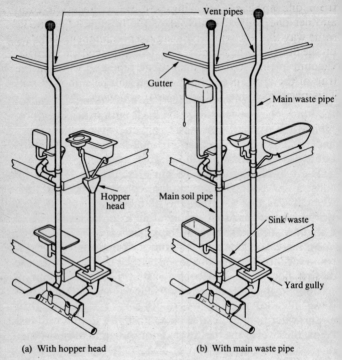

(a) With hopper head   (b) With main waste pipe

Fig. 41   Two-pipe drainage systems

open-ended to above eaves level in exactly the same way
as a main soil and vent pipe.

Two pipe drainage systems of this kind worked well
enough as far as two-storey buildings were concerned.
Problems arose however when it was necessary to provide
above ground drainage for multi-storey flats, offices and
hotels.

If two or more branch waste pipes – whether from
lavatories, baths or basins – are taken to connect to a
common waste pipe, there is every danger that waste water

from one appliance, flowing past the connection from another one, will suck air out of the branch waste in the same way that the wind, blowing over a chimney, sucks air up the flue from the room below. This will produce a partial vacuum in the second branch waste pipe and water in the trap of the appliance connected to it will be siphoned out. This phenomenon is called *induced siphonage*.

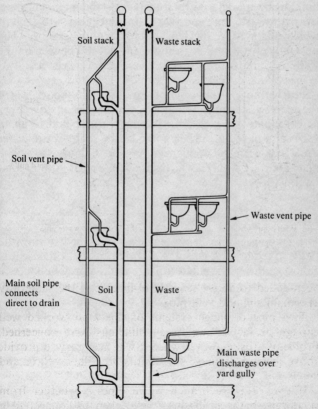

Fig. 42 Two-pipe drainage from a multi-storey building

To prevent this the trap of each appliance had to be ventilated by a branch vent pipe connected to the upper part of the outlet of the trap. These branch vent pipes were connected to a main vent pipe which either terminated open-ended above eaves level or could be taken back to connect to the main waste stack at least 3 ft above the level of the highest waste connection to it. Thus, the use of the two pipe system meant that the external walls of multi-storey buildings were defaced by four pipes – the main soil-pipe, the main waste pipe and two main vent pipes – to each of which might be connected any number of branches. The unsightly effects of this arrangement can still be seen on the walls of multi-storey buildings erected between the two World Wars.

Furthermore a blockage of the yard gully above which the main waste pipe discharged, caused perhaps by paper or cardboard blowing onto the gully grid, could result in yards and paved areas being flooded by soapy waste water while the users of appliances on upper floors would be quite unaware of the havoc that they were creating at ground level.

The one-pipe drainage system

The one-pipe drainage system was developed during the 1930s to reduce at least some of these problems. Its use was almost exclusively limited to large multi-storey buildings.

With a one-pipe system the distinction between 'soil' and 'waste' appliances was abandoned. All wastes were connected to one main soil and waste pipe, at least 10 cm (4 in) in internal diameter, which terminated open-ended above eaves level.

To guard against induced siphonage all baths, basins, bidets and sinks connected to it had to have deep seal (75 mm/3 in) traps and the trap of each appliance had to be ventilated into a main vent pipe. As before this could either terminate open-ended above eaves level or could be

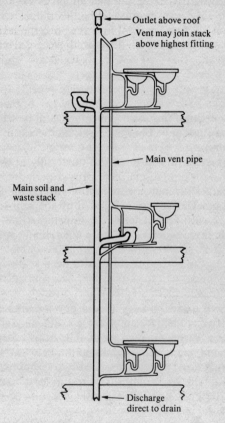

Outlet above roof

Vent may join stack above highest fitting

Main vent pipe

Main soil and waste stack

Discharge direct to drain

Fig. 43   One-pipe drainage from a multi-storey building

taken to connect to the main soil and waste pipe at a point
at least 3 ft above the highest connection to it.

This at least removed the problem of the flooding gully
and reduced to two the number of pipes climbing from
ground to roof level against the walls of the building. In
many cases, in fact, the soil and vent pipes were taken to
ground level via internal ducts and were quite invisible
externally.

Single stack drainage

During the early post-war years it was realised that the
dangers of siphonage could be overcome without the need
for an elaborate system of trap ventilation. The single stack
drainage system, that is in almost universal use today, was
the result.

Single stack drainage eliminates ventilation of branch
waste pipes. To be successful it must be constructed in
accordance with very strict design considerations.

As with a one-pipe system the main soil and waste stack
must be at least 10 cm (4 in) in internal diameter and all
waste fittings must be provided with deep seal traps.

Fittings must, as far as possible, be grouped round the
main drainage stack to give short branch waste pipes.
Gradients of branch wastes should be very slight.

These two considerations are particularly important
where basin wastes are concerned. The branch basin waste
will be of only 3 cm (1¼ in) internal diameter. It will therefore fill with waste water when the basin is emptied and the
trap seal could be broken as a result of self-siphonage.
Unlike a bath or sink there is little 'run-off' to reseal the
trap after the basin has emptied.

This branch waste must therefore be laid to a minimal
gradient and should preferably be no longer than 1.70 metres
(5 ft 6 in). If a greater length is unavoidable it may be possible
(depending upon the requirements of the local authority's
building inspector) to provide a patent anti-siphon trap or,

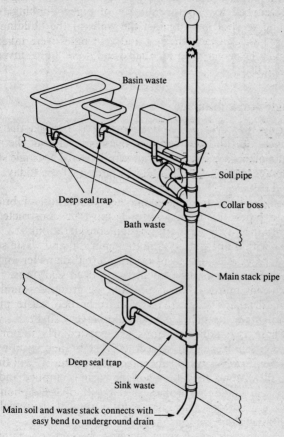

Fig. 44   Single stack drainage

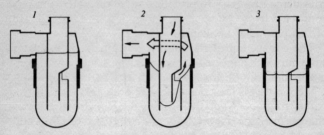

1: Shows the Barvac trap under normal operating conditions with full water seal.

2: When subjected to severe siphonage conditions the Barvac automatic hydraulic action allows air through the by-pass tube without any major loss of water.

3: When normal conditions return the remaining water falls back to re-seal the trap

Fig. 45   The *Barvac* antisiphon trap

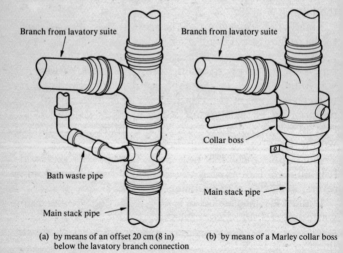

(a) by means of an offset 20 cm (8 in) below the lavatory branch connection

(b) by means of a Marley collar boss

Fig. 46   Two ways of connecting the bath waste to a single stack

perhaps, to use a 4 cm (1½ in) waste pipe. But it may be considered necessary to provide a small diameter vent pipe, connected to the main stack at high level, to vent the trap of this particular appliance.

The connection of the branch waste from any lavatory suite must be connected to the main stack by a joint 'swept' in the direction of flow.

Steps must also be taken to prevent the waste outlets from baths, basins or sinks being fouled or obstructed by discharges from the lavatory suite. To avoid this, no connection should be made to the main stack within 20 cm (8 in) of the centre point of the connection of a branch from a lavatory suite. This can present difficulties, particularly where bath wastes are concerned. These can be overcome by using a patent *Marley* collar-boss which effectively prevents the fouling of branch waste outlets while permitting them to discharge into the main stack at the same level as the lavatory outlet.

Finally the connection of the main soil and waste stack to the underground drain must be made with an *easy bend*.

Single stack drainage was originally intended, like the one-pipe system, for use in multi-storey buildings. However it was found to be equally suitable for domestic dwellings. Its universal adoption for this purpose was hastened by the requirement of the Building Regulations of the mid-1960s that all main soil and waste stacks should be contained within the fabric of the buildings that they served.

This requirement has now been substantially modified. It is unlikely though that this modification will result in a reversion to two pipe drainage.

It must be mentioned here that although ground floor waste outlets *can* be connected to the main soil and waste stack of a single stack system it is more usual to deal with them in the traditional way. Ground floor lavatory suites usually have an S trap outlet that is connected directly to a branch underground drain. Ground-floor sink wastes are taken to discharge over a yard gully.

A further, very sensible, requirement of the Building Regulations is that waste pipes discharging over gullies must do so above the level of the water in the gully but *below* the gully grid. This prevents gullies from flooding as a result of blown paper or leaves blocking the gully grid. It ensures too that the full force of the discharging waste water is available to flush and cleanse the gully.

Side and back inlet gullies are available to permit waste water to enter below the grid. There are also slotted gully grids that make it possible to adapt existing gullies to below-grid discharge.

With one-pipe and single stack drainage systems the requirement that lavatory compartments should have means of permanent ventilation and that an 'intervening ventilated space' must separate them from habitable rooms is retained though for aesthetic, rather than health, reasons.

Blocked waste pipes

No system of above-ground drainage is so perfect that a blocked waste pipe will never occur. The plug is pulled from the waste outlet of a sink or basin – and nothing happens. The appliance remains full of greasy, or soapy, water.

Every householder should have in his possession a force cup or sink waste plunger to deal with this kind of emergency. A force cup is a hemisphere of rubber or plastic, usually mounted on a wooden handle, and can be bought from any d-i-y shop or household store.

To use a force cup place the rubber hemisphere squarely over the waste outlet of the sink or basin. Hold a damp cloth firmly against the overflow outlet while, with the other hand, plunging the handle of the force cup sharply downwards three or four times.

Since water cannot be compressed the effect of this is to convert the column of water in the waste outlet into a ram to dislodge the blockage. The overflow outlet is blocked to prevent this force from being dissipated.

If, at first, your efforts appear to be having no effect, continue plunging. You may be moving the obstruction further and further along the branch waste pipe until eventually it is forced out into the gully or the main waste pipe.

Where plunging proves totally ineffective it is probable that some solid object has become wedged in the trap itself. To clear it you will need to gain access to the trap.

Place a bucket under the trap before you attempt this. Simple U traps have an access cap near their base that can be unscrewed and removed. The entire lower part of a bottle trap can be unscrewed, usually by hand, to give access.

Probe through into the trap and waste pipe with a length of expanding curtain wire or other similarly flexible wire.

Partial blockages

If a sink empties more slowly than it has in the past it is probable that a build-up of grease in the waste pipe is causing a partial blockage. This can be cleared with the use of a proprietary drain cleaning chemical such as *Drain-free*. These chemicals have a caustic soda base, and should be handled with extreme care and used in strict accordance with the manufacturer's instructions. Keep in a locked cupboard well away from children. Smear Vaseline over the chromium plated waste outlet before introducing them into the waste pipe. Avoid getting them on your hands.

A partial blockage of the outlet of a bath or wash basin may be caused by a build-up of soap debris on the walls of the waste pipe. This too can be cleared with drain-clearing chemicals.

Basin wastes often become partially blocked by hair trapped by the outlet grid. Probing through the grid with a piece of wire will often reveal large quantities of hair. If this is difficult to remove, unscrew the large nut connecting the waste outlet to the inlet of the trap. You will then be able to push the trap on one side and pull it out from beneath.

Drain smells

Drain smells in a bathroom or bedroom may be caused by a number of factors.

If you have a two-pipe drainage system, do the bathroom wastes discharge into a rain-water hopper head in the vicinity of the bathroom or bedroom window? If so, cleanse it and flush through the downpipe with a hot solution of washing soda. Cleanse and flush the yard gully into which the downpipe discharges.

Is the water in the trap of either the bath or the wash basin siphoning out? Looking downwards through the plug hole you should be able to see the surface of the water in the trap just below. You should, in any case, be able to smell whether this is the source of the trouble. If it is, it may be wise to seek the on-the-spot advice of the Environmental Health Officer of your local authority. Trap ventilation or the provision of a patent anti-siphon trap is likely to be the solution.

Does the smell emanate from the built-in overflow of the wash basin? This can be very difficult to clean effectively. Smells from this source can usually be eliminated by spooning drain-cleaning chemicals directly into the overflow inlet.

Another possible source of smell is a leakage from the connection between the lavatory pan outlet and the branch drain or soil-pipe. A means of remedying leaks of this kind is discussed in Chapter 5.

Are you quite sure that the smell comes from the drains? A badly rinsed face flannel will soon begin to smell offensively. Some plastic electric fittings produce a particularly objectionable 'fishy' smell if arcing causes them to heat up. A smell from this source will become evident within a few minutes of the electric light, or other electric appliance, being switched on.

Rain water drainage

Collection and disposal of rain water from roofs is an important aspect of above-ground drainage. Inadequate, badly fitted or leaking rain water systems are a common cause of dampness in walls which, in turn, can produce wallpaper stains, dry rot and other structural troubles.

Design considerations

Despite the fact that annual rainfall in some parts of Britain is three or four times as high as it is in others the same design factors can be used for the whole of the British Isles. This is because rain-water system design depends on the intensity of rainfall likely to be experienced at any one time.

Intensity varies little, if at all, throughout the country. A summer thunderstorm in north-east Essex will have a similar intensity to one in the Lake District despite the great difference in annual rainfall between the two areas.

Nevertheless no rain-water system can be designed to cope with any downpour that could possibly occur. As a basis for design, a maximum intensity of 3 in per hour is assumed. In any particular locality rainfall of this intensity may be expected for about five minutes over a two year period. Rainfall with an intensity of 4 in per hour can be expected for about five minutes once in five years or for ten minutes once in nineteen years. No great harm will result from gutters overflowing briefly on those rare occasions.

Using these design factors a half-round gutter with a radius of 6 cm ($2\frac{1}{4}$ in) laid to a fall of 1 in 600 – 1 in (25 mm) in 50 ft (15 m) – draining to a downpipe 65 mm ($2\frac{1}{2}$ in) in diameter will cope with a roof area of 110 square metres (125 square feet) if the downpipe is situated in the centre of the gutter and 60 square metres (630 square feet) if the downpipe is at the end.

The roofs of a pair of small terraced or semi-detached houses could therefore be drained by a system of this kind having one downpipe at the front and one at the rear of the properties. A similar system, with two downpipes, would be adequate for the average detached suburban home though larger properties might require one or more extra downpipes.

Materials used

Cast iron

Pre-war homes were nearly always provided with cast iron gutters and downpipes. The gutters, either of half-round or of *ogee* pattern, were joined by bedding the spigot end of one length into the socket end of another on a bed of putty. The joint was then secured with a gutter bolt. Systems of this kind require regular painting to protect them from corrosion.

Leakages may occur in a cast iron system as a result of a length of pipe or guttering splitting, following exposure to frost or corrosion or, more commonly, as a result of a failure of the joints between lengths of guttering.

To renew a leaking joint remove the gutter bolt. It will usually be found easier to saw through it flush with the gutter rather than to attempt to unscrew it. Clean the gutter's socket and spigot. Remove rust, and paint as necessary. Then re-bed on a non-setting mastic filler such as *Plumbers Mait*, finally securing the joint with a new gutter bolt.

Cracks can be repaired, after thorough cleansing and removal of rust from the affected area, with an epoxy resin filler.

Where a cast iron rain water system is found to be badly cracked and corroded it is usually better to dismantle it and to replace it with a modern plastic system.

Asbestos cement

Asbestos cement gutters and downpipes enjoyed a brief vogue after the Second World War. They do not corrode and need no decoration. They are however rather heavy and

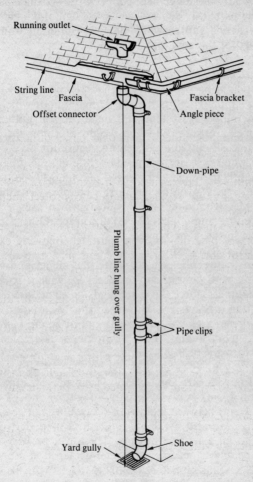

Fig. 47   A uPVC rain water system

clumsy in appearance and are very subject to accidental damage. An asbestos cement gutter can be broken if a ladder is placed carelessly against it and accidental contact with, for instance, a garden wheel barrow, can easily damage a downpipe of this material.

Unplasticised polyvinyl chloride

Unplasticised polyvinyl chloride (sometimes called uPVC, PVC or simply vinyl) is the plastic material of which most modern rainwater systems are made. uPVC gutters and downpipes are tough, light in weight and non-corroding. They are easily fitted and need no decoration.

Rain-water systems produced by different manufacturers vary slightly in design and method of fixing. The gutter and downpipe sizes given earlier in this chapter were those of the OSMA Roundline $4\frac{1}{2}:2\frac{1}{2}$ system which uses a 11 cm ($4\frac{1}{2}$ in) gutter and a 7 cm ($2\frac{1}{2}$ in) downpipe. Other sizes are available in the same range.

OSMA systems can be installed with a basic tool-kit. Junctions between lengths of guttering, angles and outlets are made with simple clip-on brackets which, to ensure water-tightness, force the gutter ends on to neoprene pads. Supporting brackets are screwed to the fascia board of the roof at 1 metre (1 yard) intervals and at 50 cm (1 ft 6 in) distance from angles. The downpipe is assembled with straightforward push-on joints.

With any uPVC system provision has to be made for thermal movement. The OSMA system provides for a gap of 6 mm ($\frac{1}{4}$ in) between lengths of gutter and of about 12 mm ($\frac{1}{2}$ in) between lengths of downpipe.

Fittings are also available to connect a modern uPVC system to an existing iron one. These can be very useful where the owner/occupier of a semi-detached house wishes to modernise his home but is unable to persuade his neighbour to follow his example.

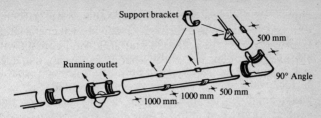

Place fingers inside gutter and both thumbs on outside of bracket, high up towards front nib. The gutter must be pulled down with the fingers while at the same time the bracket is pushed forward with the thumbs until it snaps over the gutter.

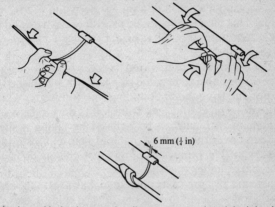

Completed assembly showing expansion allowance necessary in each fascia bracket

Fig. 48   Detail of fixing an OSMA roundline uPVC gutter

Rain water disposal

Rain water disposal does not, strictly speaking, come within the scope of 'above ground' drainage but it is convenient to consider it at this point.

Many rain-water downpipes discharge over yard gullies connected to the main drainage system of the house. From the point of view of the sewerage authority this is not a satisfactory arrangement. It means that sewers are overloaded and liable to flood during periods of heavy rain. It means too that the composition and quality of sewage reaching the sewage treatment works varies continually according to weather conditions.

For this reason many urban authorities provide separate surface water sewers for the reception of water from roofs and from highway drains. These usually discharge their contents, untreated, into a convenient river or stream.

Where there is no separate surface water sewer the authority may require rain water to discharge into a soak-away constructed on the householder's own land. A typical soakaway might consist of a pit 5 ft deep and 4 ft from side to side. This pit would be filled in with brick-bats and other rubble to within about 1 ft of the surface and the surface soil then reinstated.

After a soakaway of this kind has been in use for a number of years it is liable to become clogged with silt washed from the roof and from the surrounding soil. Rainwater will back up the downpipes and leak from joints during heavy rain.

Clogging with silt can be reduced by laying a sheet of polythene over the infilling material before reinstating the surface soil. The rain water drain must, of course, be taken to discharge under this sheet.

There are also pre-cast concrete soakaways on the market. These are less likely to become clogged and, if they do, it is a relatively simple task to remove the manhole cover that gives access to them and to dig out the silt.

9

Underground Drainage

Above ground, the only visible signs of the underground drainage system of a house are the cast iron covers of the drain's inspection chambers or manholes. Every drainage system is likely to have at least two of these. One will be situated near the point at which the main soil and vent pipe disappears into the ground. The other will be near the front boundary of the property – in the drive-way or the front garden perhaps. What lies beneath those covers and under the ground between them will depend largely upon the age of the property.

The drains of pre-war houses

Pre-war drains could be made of glazed stoneware or of heavy iron pipes suitably protected against corrosion. The use of iron pipes was restricted almost exclusively to good class commercial and industrial work.

Domestic drains were usually constructed of 2 ft long glazed stoneware drain pipes 4 in in internal diameter. They were laid to what it was hoped would be a 'self-cleansing' fall on a 6 in thick bed of concrete. The purpose of the concrete base was to protect the drains from the effect of any ground settlement that might take place.

To give the drainage system greater stability it was usual to 'haunch' the concrete to the top of the pipe. Where drains of this kind passed under buildings, local authority bye-laws required that they should be totally encased in a 6 in thickness of concrete.

Stoneware drain pipes were joined to each other by means either of a neat cement joint or – more usually – a joint made of a mixture of two parts of cement to one of sand. Before the joint was made a rope grummet was pushed over the spigot end of the pipe and was caulked down hard into the socket of the pipe to which it was to be connected. This was to prevent jointing material from being extruded into the drain – a common cause of drain blockages.

It had been discovered by experiment that drains were self-cleansing when drainage flowed through them at the speed of 3 ft per second. A rule-of-thumb that was adopted universally in the pre-war years was to lay 4 in drains at a gradient of 1 in 40 (3 in in 10 ft) and 6 in drains – used where several properties were drained in combination – at a gradient of 1 in 60.

This easily remembered formula generally gave satisfactory results though, with a well-laid drain, the rate of flow would be in excess of 3 ft per second. Drain blockages can arise as a result of too great, as well as a result of too small, a fall. If the gradient is too steep liquids will flow on in advance of solid materials. These latter may remain in the drain to build up into a blockage.

Drains were laid in straight lines and inspection chambers had to be provided at all junctions and major changes of direction. The purpose of the inspection chambers was to permit every part of the drain to be accessible to drain rods when it was necessary to clear an obstruction.

Drain inspection chambers

Pre-war inspection chambers were built in either 4 in or 9 in brickwork, depending upon the depth of the chamber. The

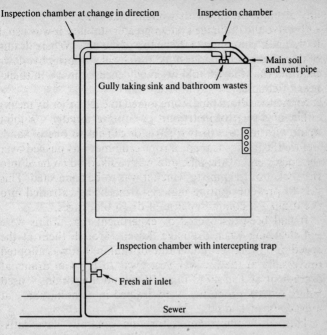

Fig. 49 Typical pre-war suburban drainage system

chamber had a concrete base and walls were usually rendered internally with sand and cement to ensure watertightness.

The main drain ran through the chamber in a stoneware half-channel built into its concrete base and branches joined it, in the direction of the flow of the drain, via stoneware three-quarter section bends designed to prevent splashing. As a further precaution against the fouling of the inside of the inspection chamber, the concrete round the channels was sloped upwards to the chamber walls and rendered in sand and cement to a smooth surface.

The intercepting trap and fresh air inlet

It was usual in pre-war drainage systems to provide an intercepting or disconnecting trap in the final inspection chamber of the household drain before its connection to the sewer. The purpose of this trap was to prevent sewer gases entering the domestic drainage system.

The necessity for intercepting traps had been questioned by sanitary engineers as early as 1911. Sewer gases are no

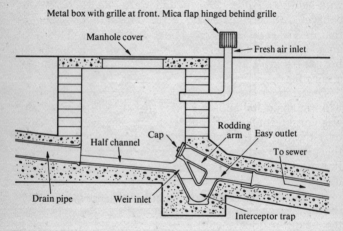

Fig. 50   Drain inspection chamber with intercepting trap and fresh air inlet

more noxious than drain smells. Furthermore if the intercepting trap were omitted from the drains of a group of houses, the sewer that served them would be perfectly adequately ventilated by the open-ended soil and vent pipe of each house.

The intercepting trap was a common cause of drain blockages. My own experience suggests that these traps are the site of something like 90% of all blockages of pre-war drainage systems.

To reduce the risk of blockages intercepting traps were constructed with an 'easy' outlet and a weir inlet 2 in deep. It was vitally important that, when the drain was laid, the trap should be set dead level on its concrete base. Intercepting traps were provided with rodding arms to permit the clearance of any blockage that might occur between the trap and the sewer. The inlet of this rodding arm was closed with a stoneware stopper.

The intercepting inspection chamber was usually provided with a *fresh air inlet*. A 4 in stoneware *knuckle bend* was connected to the upper part of the inspection chamber so that its socket projected just above ground level. The fresh air inlet was cemented into this socket. It consisted of a metal box with a grille at the front. Behind the grille was hinged a mica flap.

The fresh air inlet was intended to ensure, in conjunction with the main soil and vent pipe, that the drainage system was constantly flushed with fresh air. The wind, blowing over the top of the main vent pipe, would suck air from this pipe thus reducing air pressure in the drainage system. Pressure would then be restored by air flowing past the mica flap into the fresh air inlet.

Should there be a reverse flow of air through the drain the mica flap would be forced against its seating to prevent the escape of drain smells.

A major failing of fresh air inlets was their susceptibility to accidental damage and vandalism. This can be confirmed by a stroll round any residential estate developed during the 1920s or 1930s. Fresh air inlets will be found broken away from the drain pipe to which they were fitted. Others will have their grilles and mica flaps broken or missing. In yet others the mica flap will be jammed either open or shut.

A number of householders may have removed the metal box and sealed off the inlet. They will have discovered that it served little useful purpose and was a common source of unwelcome smells.

Post-war underground drainage

Although 'post-war' is a convenient description of the period during which radical changes in drainage construction took place, it should be realised that these changes did not take place overnight. The drainage systems of many houses built in the 1950s and even the 1960s are exactly the same as those already described.

Changes relate, in fact, rather to materials and the ways in which they are used than to the basic principles of drainage design.

Drains today must still be laid in straight lines to a self-cleansing gradient. Means of access – usually inspection chambers – must still be provided at major changes of direction and at points where branches connect to the main drain.

Drains are however no longer constructed of cement jointed stoneware pipes, with each joint a potential point of leakage. These have been replaced by long lengths of pitch fibre or uPVC pipe connected by simple push-fit ring-seal joints. These materials are capable of accommodating slight subsoil settlement so the 6 in concrete raft on which stoneware drains were laid has been discarded. Trench beds must still be carefully prepared though. In some cases it may be necessary to bring on-site suitable material on which to lay the drains.

Another effect of the new materials has been to make it possible to reduce drain gradients. Falls of 1 in 60 or 1 in 70 have been found to give perfectly satisfactory results.

It should perhaps be mentioned that the fact that the sewer may be laid deep under the highway should not affect the gradient of the house drains. These are laid at the suggested falls to the final inspection chamber before connection to the sewer and then fall steeply to the sewer itself.

Modern inspection chamber construction

Although some uPVC drainage systems are supplied complete with *rodding points* that virtually eliminate the need for inspection chambers, these still provide the most common means of gaining access to underground drains.

They may, in fact, still be built in brickwork in the traditional style with stoneware half channels to which pitch fibre or uPVC pipes can be connected. This brickwork should *not* be rendered internally. It has been found that

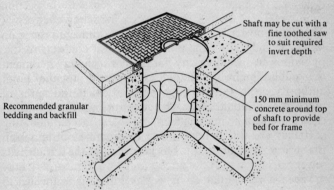

Shaft may be cut with a fine toothed saw to suit required invert depth

150 mm minimum concrete around top of shaft to provide bed for frame

Recommended granular bedding and backfill

4D.857 Inspection Chamber Invert depth 570 mm maximum

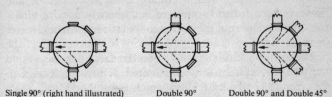

Single 90° (right hand illustrated)　　　Double 90°　　　Double 90° and Double 45°

Some alternative arrangements of base

Fig. 51　An OSMA drain GRP inspection chamber

internal rendering tends to flake off the walls of inspection chambers creating drain blockages. Rendering should be to the exterior of the walls and should be done before the soil is reinstated after the construction of the chamber.

Inspection chambers may also be made of pre-cast concrete sections or of fibre-glass reinforced plastic (GRP). GRP chambers are circular in plan and are constructed, in one piece, with a variety of alternative bases capable of taking up to four branch drains into a central half-channel. They may be cut, with a fine toothed saw, to the required depth.

Some legal points

The house-owner is responsible for the drainage system of his home right up to the point at which it connects to the public sewer, including any part of it which may lie under the public highway. Before complaining to the local authority or to the sewerage authority that the sewer is blocked, you should check whether or not your neighbours' drains are also giving trouble.

If they are still running freely there will be little doubt that it is *your* drain that is blocked and not the authority's sewer.

When a residential estate is developed, it is the usual practice for the estate developer to arrange for up to about a dozen houses to be drained 'in common'. This cuts the cost of drain laying and of connection to the sewer. It can however be a fruitful source of neighbourly discord when the 'common' part of the drainage system needs to be cleared or maintained.

The legal position depends upon the exact date on which the drainage system was constructed. If this was before 1 October 1937 (the date upon which the Public Health Act 1936 came into force) this 'combined drain' is designated as a 'public sewer'. The sewerage authority is responsible for its repair, maintenance and 'cleansing'. The authority

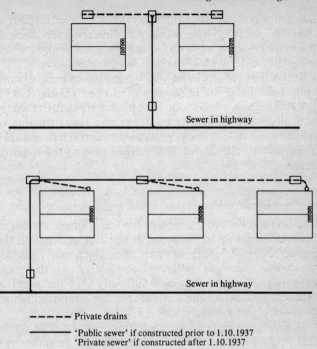

----- Private drains

———— 'Public sewer' if constructed prior to 1.10.1937
'Private sewer' if constructed after 1.10.1937

Fig. 52 'Private sewers' and 'public sewers'

can however recover the cost of repair and maintenance –
but not of cleansing – from the owners of the properties
concerned.

'Cleansing' is usually assumed to include the clearance
of blockages so, except in the case of recurring blockages
caused by a structural defect, users of such a 'public sewer'
can expect to have this work done free of charge.

Combined drains constructed after 1 October 1937 are
designated as 'private sewers' for which the owners of the
properties draining into it are wholly responsible.

If you think that your home may be drained in combination with others see if there is any references to this fact in your house deeds. Call at the offices of the local authority and ask to see the drainage plans. If there is any doubt about the situation ask if the authority's Environmental Health Officer will call and advise you. Once you have established the position make sure that the information is passed on to your neighbours.

It cannot be stressed too strongly that responsibility for the clearance, maintenance and repair of combined drains should be established *before* trouble arises. It frequently happens that, when a combined drain is blocked, the garden of the householder nearest to the sewer is flooded while residents of properties at a slightly higher level continue to pour sewage into the drainage system in blissful ignorance of the fact that the crisis down the road, has anything whatsoever to do with them.

Clearing blocked drains

There are several ways in which a householder may become aware that his drainage system has become blocked. When a lavatory is flushed water may rise almost to the flushing rim of the pan and then slowly subside. Water may ooze from under the rim of an inspection chamber cover or a yard gully may flood. In the latter event, first try removing the gully grid. It could simply have become clogged with leaves or blown paper.

Raise the covers of the inspection chambers to establish the position of the blockage. To do this, run the blade of a spade round the edge of the lid to remove accumulated silt. It should then be possible to raise the cover by the two handles provided. If, over the years, these have rusted away, force the spade blade between the lid and its frame and use it as a lever to raise it.

Having removed the inspection chamber covers you will be able to decide the position of the blockage. You will need a set of drain rods or sweeps rods and perhaps a rubber drain plunger to clear it.

If all the inspection chambers are flooded and your drainage system has an intercepting trap you can be pretty certain that it will be this trap that is blocked. Screw two lengths of drain rod together and screw the drain plunger – a 10 cm (4 in) diameter rubber disc – on to the end.

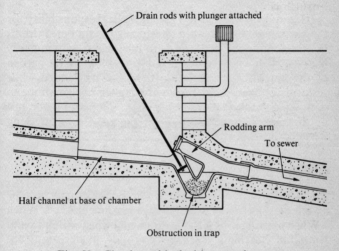

Fig. 53   Clearing a blocked intercepting trap

Lower the plunger into the inspection chamber and feel for the half-channel at the bottom. Move the plunger along the channel towards the sewer until you can feel the drop into the intercepting trap. Plunge down sharply three or four times. There will probably be a gurgle and water level in the chamber will begin to fall as water flows freely through the trap.

In an emergency an old fashioned kitchen mop or even a bundle of rags tied *securely* to a broom handle can be pressed into service as a drain plunger.

If one inspection chamber is flooded and another, nearer

to the sewer, is clear, then the blockage must exist between these two chambers. Screw two or three lengths of drain rod together and lower the end into the flooded chamber. Feel for the half-channel at its base and push the rods along this and into the drain in the direction of the empty chamber. Screw on more rods as may be necessary and continue to thrust along the drain until you encounter and dislodge the obstruction.

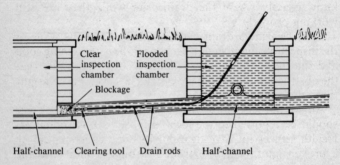

Fig. 54   Clearing a blocked drain

When pushing rods into the drain, and when withdrawing them after clearing a blockage, you may find it helpful to twist them clockwise. Never – however firmly they may appear to be stuck – be tempted to twist in the other direction. If you do they will unscrew and some will be left in the drain.

A drain blockage between the intercepting trap and the sewer is, fortunately, a rare occurrence. It is indicated by a flooded intercepting chamber that cannot be cleared by plunging.

To clear a blockage in this position you will first need to remove the stopper from the rodding arm. This will be submerged under several feet of sewage. Fortunately the stopper is provided with a round knob and it is usually possible to get a rod behind this knob to remove it. Having done this, push the rod through the arm into the drain beyond it. The chances are that you'll find the obstruction to be at the actual connection between the drain and the sewer.

Needless to say you will find it much easier to deal with blocked drains and flooded inspection chambers if you are already thoroughly familiar with the layout of the drainage system. Raise the manhole covers and make sure that you know exactly where the drains run while they are still working properly.

After clearing a drain blockage flush down the walls of the inspection chambers with a hot solution of washing soda and run the taps for half an hour or so to make sure that the obstruction that you have cleared has been flushed right through to the sewer.

Any drain may block occasionally. Recurring blockages should be reported to the local authority's Environmental Health Officer for investigation. Possible causes could be a broken joint, cement extruded through a joint into the drain or tree roots which will unfailingly find any slight break through which they can grow to cause an obstruction.

Drain smells

An unpleasant smell that persists near the front boundary of an older property will usually be found to be caused by a partial blockage of the intercepting trap. Often it will first be noticed shortly after a period of heavy rain.

During heavy rain the sewer will surcharge and back pressure may push the stopper of the rodding arm out of its socket. It will fall into the trap immediately below to cause a blockage.

Unlike most drain blockages it is one that may remain undiscovered by the householder for weeks or even months. Sewage will rise in the inspection chamber until it reaches the level of the rodding arm and will then flow down this arm into the sewer. In the meantime the sewage in the

bottom of the chamber will become more and more foul
and sewer gases will be able to escape near ground level
either from under the inspection chamber cover or from the
fresh air inlet.

The remedy is, of course, to extract the stopper from the
trap and to cleanse the inspection chamber thoroughly.
Where this has occurred once it could occur again. To
prevent a repetition, discard the stopper and cement a disc
of glass or slate lightly into the socket of the rodding arm.
This will not be forced out by back pressure. In the event
of a drain blockage between the intercepting trap and the
sewer the disc can be broken with a crowbar and replaced
after the drain has been cleared.

Drain smells can sometimes escape from under the rim
of a badly fitting manhole cover. This can be prevented by
running grease into the slot of the frame into which the cover
fits.

Where waste pipes discharge over the grid of a yard gully
these can become fouled and smelly. Grids need regular
cleansing. If you have a solid fuel room heater or boiler you
can cleanse and sterilise metal (but not plastic!) grids very
effectively by placing them on the fire.

A better, and more permanent, solution to this problem
is to replace the existing grid with a slotted one of the kind
referred to in the previous chapter. Extend the waste pipe
through this slot to discharge beneath it. This will not only
prevent fouling of the grid but will ensure that the gully itself
is regularly cleansed and flushed through by the waste
discharges.

As with drain smells in the bathroom, consider the possibility that the drains are *not* responsible. A leaking gas
pipe can produce a smell resembling that of sewer gas. The
unexplained death of shrubs or part of a boundary hedge
in the front garden can be an indication of a gas leak.

Leaking drains

Constantly recurring blockages are the commonest sign of a defective drain. Other possible indicators are a persistent patch of dampness on the wall of a basement or on a driveway or the presence of rats in the garden and rat droppings on the benching of drain inspection chambers.

If you have any reason to believe that the drains of your house may be defective, ask the local authority's Environmental Health Officer to call and make an inspection. There are a number of tests that he can apply to the drainage system. One of the most effective is to stopper off the suspected section of drain and introduce smoke into it under low pressure. A test of this kind will establish beyond doubt whether or not the drain is defective and, if it is, will indicate the position of the defect.

The Environmental Health Officer will also make arrangements for any rats that may have infested the drainage system to be destroyed.

10

Rural Water Supply and Drainage Problems

Those of us who live in towns tend to take our water supply and sewerage services for granted. We turn on the taps and flush the lavatory suite without a thought of the ultimate source of our water supply or the destination of our household's drainage. Move to a cottage in the country and you may find that these are matters that force themselves to your attention.

Rural water supplies

Unless you move to a very remote and isolated rural home, water supply is unlikely to present a serious problem. A safe water supply is a primary necessity of life and water authorities have been very active in extending mains water supply to villages and rural communities throughout the country.

There will however always be some isolated properties to which it would be uneconomical or impracticable to take a mains supply. Such properties usually have to rely upon a well within their own grounds.

Wells are classified as either *deep* or *shallow*. This classification does not relate to the actual depth of the well but to the source of the water that it taps.

Water soaking through the soil may be trapped by the first layer of impervious 'rock' that it encounters. This could be a layer of clay no more than 10 ft below the surface. Some water will succeed in penetrating this barrier, through a fault or through an outcrop of porous rock some distance away. It will soak through the next permeable stratum, which might perhaps be chalk or limestone, and will finally be trapped by a further layer of impervious rock at a greater

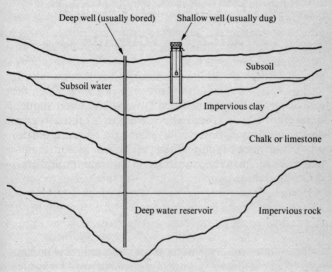

Fig. 55   Deep and shallow wells

depth. A shallow well is one that taps the subsoil water above the first impervious stratum of the earth. A deep well is one that taps water below the first impervious stratum. Shallow wells are usually dug out and lined with either brickwork or concrete rings. Deep wells are usually bored and the water drawn up through a metal pipe of relatively small diameter.

The distinction is thus sometimes made between *dug wells* and *bore wells*. This is not a reliable classification. Dug wells do sometimes tap water below the first impervious stratum. Similarly a bore well may tap only subsoil water.

Shallow wells are *always* suspect as a source of water supply. The water upon which they draw may be contaminated from nearby fields or farmyards, by defective drains or by rural sewage treatment systems. They are also unreliable and may dry up during periods of prolonged drought.

Water from deep wells is much more likely to be safe for drinking. Any dangerous organisms are likely to have been filtered out during the journey to its underground reservoir.

Even deep wells can become contaminated though. Anyone relying upon well water should ask the Environmental Health Officer of the local authority to take samples for bacteriological examination at regular intervals. Even if the well is seriously contaminated it is very unlikely that this examination will reveal the presence of actual disease organisms. It may however reveal the presence of, in themselves, harmless organisms that are normally found in the intestines of man and animals. These are indicators of serious pollution.

If such organisms are present the Health Officer will advise on the best course of action to take. It could be that the well will need to be relined. Alternatively some kind of filtration or chemical sterilisation may be desirable.

If the water supply is bacteriologically pure it will almost certainly be safe for adult use. This will not necessarily mean that it is safe for very young children to drink.

The normal processes of the decomposition of organic matter result in the breakdown of complex ammoniacal compounds into nitrites and nitrates. These chemicals may also enter a source of water supply as a result of the agricultural use of chemical fertilisers. If water containing substantial amounts of nitrites or nitrates is used to mix up a baby's feed, oxygen may be extracted from the baby's blood to produce the serious and sometimes fatal 'blue baby' condition.

Local authorities are notified by the Health Authority when a baby is expected in a home served by a well water supply. The Health Officer then takes samples of the water for chemical analysis as well as for bacteriological examination. If there is found to be a high nitrite or nitrate content the local authority will arrange for an alternative supply – perhaps the delivery of a few gallons of water a day – during the first critical months of the baby's life.

Safe storage of well water

The days of water extraction by bucket and windlass are, except for the publishers of picture postcards, happily past.

In most cases water is drawn from the well by means of a float-controlled submersible electric pump and is stored for use in a large, high level storage cistern, usually within the roof space of the house. This cistern will supply not only the hot water system and the bathroom plumbing fittings but also the cold tap over the kitchen sink – the household's source of water for drinking and cooking.

It is therefore doubly important that this cistern should be effectively covered to protect it from the risk of contamination. It should be drained annually and thoroughly cleansed with a suitable hypochlorite solution.

Rural drainage

The provision of sewerage systems in rural areas has not proceeded at the same pace as the laying of water mains. Many rural communities, as well as isolated cottages, have no access to a sewer. In other areas, although a sewer may be available, some house owners will have preferred to avoid the cost of connection to it and will have continued to rely upon septic tank or cesspool drainage. Unless 'a nuisance' can be proved local authorities' powers to compel connection are strictly limited.

When purchasing a modernised cottage, or a cottage with

the intention of converting it to a modern home, it should be remembered that a drainage system that was perfectly satisfactory for the original occupiers may be quite incapable of coping with wastes from present day bathroom and kitchen plumbing equipment.

Less than half a century ago the only plumbing fittings likely to be found in a rural cottage were a shallow sink and – perhaps – a lavatory in an outhouse. Automatic washing machines were unheard of. Baths and wash-hand basins were a rarity.

Cesspools

Rural properties, where there is no sewer available, may be drained either to a cesspool or to a septic tank. The difference between the two systems is not always appreciated by potential purchasers and an estate agent, eager to make a sale, is unlikely to dwell on the problems that cesspool drainage can create.

A cesspool is simply a watertight underground chamber designed to contain sewage until it can be pumped away and disposed of. It may be constructed of brickwork or it may consist of a number of precast concrete rings set on a concrete base. It may have a capacity varying from less than 500 gal to 4000 gal. The Building Regulations require that cesspools must have an actual capacity (below the level of the inlet) of 18 cubic metres. A cesspool of this size would contain about 4000 gal of sewage. Remember though that most cesspools in this country were constructed long before the advent of the Building Regulations. They are much more likely to have a capacity in the region of 1000 gal. A cesspool constructed of 54 in diameter concrete rings would have a capacity of 98 gal per foot of depth. Thus, if it were 10 ft deep *below the inlet* it would have a capacity of 980 gal.

How often would a cesspool of this capacity need to be emptied?

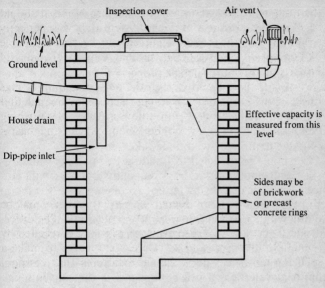

Fig. 56 Typical cesspool

Water authorities estimate a daily consumption of about 30 gal per head for domestic use. This, of course, includes water used for baths and domestic laundry as well as for personal washing, drinking, cooking, lavatory flushing and so on.

Every drop of water drawn off within the home will find its way, in one form or another, into the cesspool. Thus a family of four might expect to use, and run to waste, about 120 gal per day – or 840 gal per week. In a little over eight days a 980 gal capacity cesspool would be full.

Few cesspools need emptying quite as frequently as this because, despite the requirements of the Public Health Acts and the Building Regulations, few cesspools are really watertight. This can however be a mixed blessing. If water can leak out it can also leak in. Where subsoil water level is high a leaking cesspool can be two-thirds full of water again within an hour of its being emptied.

Cesspool emptying is an unpleasant, smelly and expensive business. Before purchasing a property with cesspool drainage estimate the capacity of the cesspool and enquire at the offices of the local authority about the cost and availability of their emptying service. If they do not themselves operate a cesspool emptying service they will put you in touch with a private contractor who undertakes this work.

Cesspools constructed to Building Regulations standard will need emptying far less frequently but this will not necessarily mean a saving in the cost of emptying. Cesspool emptying vehicles are of limited capacity. It is often possible to deal with the contents of a small pre-Regulations cesspool in one load. Several expensive journeys may be necessary to empty a 4000 gal cesspool.

The larger the cesspool, the more difficult it is to ensure that it is watertight if it is constructed in the conventional way. To get over this difficulty one-piece fibreglass reinforced plastic cesspools are manufactured. They need simply to be sunk into a pit in the earth though, to prevent subsoil water pressure from forcing them out of the ground when they are empty, some concrete work – particularly in the form of a concrete slab constructed *in situ* above the cesspool – will be necessary. The watertight qualities of these cesspools can be guaranteed.

Such cesspools usually have a capacity of 2000 gal. Two can be linked to provide the capacity required by the Building Regulations. It must be added though that local authorities have considerable power to modify the Building Regulations where they consider that this would be justified by local circumstances. Where there are plans to provide a sewer in the foreseeable future a district council might well be prepared to permit the provision of a cesspool of only 2000 gal capacity.

This is a matter about which the Building Control Officer

or Building Inspector of the district council should be consulted.

Septic tanks

A septic tank installation – at its worst – can be no more than a small, leaky cesspool. At its best it can provide an efficient and unobstrusive means of sewage purification and disposal.

Reference to the Building Regulations about septic tank construction will reveal little more than requirements that it must have a capacity of at least 2.7 cu metres (600 gal), it must be covered or fenced in, properly ventilated and so sited as to avoid the possible contamination of any source of water supply and to avoid being, 'a source of nuisance or a danger to health'. Details of siting and construction are left to the local authority and to the Area Water Authority. The latter authority is concerned because of their overall responsibility for water supplies and for the avoidance of pollution of streams and waterways.

This dependence upon *local* control in the light of local circumstances is as it should be. Safe septic tank design and installation depend upon the nature of the subsoil, the level of the subsoil water and the proximity of other properties and of possible sources of water supply. A septic tank installation that might be perfectly satisfactory for an isolated rural property could be a positive danger to health on the outskirts of a populace village.

To assist householders and builders, many local authorities publish an 'approved plan' of a septic tank installation suitable for their particular area though these plans too may need to be modified in the light of even closer local circumstances.

The septic tank itself consists of an underground chamber, which is usually rectangular, designed to retain sewage for a period of at least twenty-four hours to enable anaerobic bacteria (organisms which flourish in the absence of free

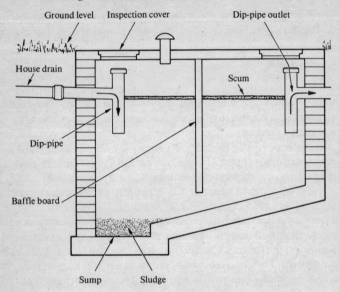

Fig. 57   A septic tank

oxygen) to carry out the initial stages of decomposition.

The inlet and outlet to the septic tank should be set to give a very slight fall and should take the form of dip-pipes passing through the shorter sides of the rectangle. These dip-pipes ensure that sewage enters and leaves the tank from points below the level of the surface of the liquid that the tank contains. To prevent a sudden heavy flow of water – from the discharge of a bath for instance – from passing straight through the tank from inlet to outlet, a *baffle board* may be provided. Alternatively the tank may be divided into two sections with a further dip-pipe linking them.

Bacteriological action within the tank has the effect of liquefying the sewage. A scum will form on the surface which will prevent the escape of unpleasant smells if undisturbed. A sludge, which must be pumped out from time to time, will sink to the bottom. A liquid effluent will remain between the sludge and the layer of scum and it is this liquid that flows from the final outlet.

The septic tank may be covered with a concrete slab provided with manholes to give access to the inlet and outlet. In this event ventilators must be provided. Alternatively the tank could be covered with heavy wooden planks. These can be removed when access is required and the spaces between the planks will allow for adequate ventilation.

Disposal of the effluent

The ultimate disposal of the effluent from the septic tank is the most critical part of septic tank installation. It is here that local knowledge is of supreme importance.

The septic tank does not 'purify' sewage. It carries out only the first phase of the processes of decomposition. The effluent will still be extremely foul and must not be permitted to flow untreated into a ditch or watercourse. How it is dealt with will depend both upon the lie of the land and the nature of the subsoil.

Where there is a fall of some 4 ft to 5 ft from the septic tank outlet to a ditch or stream the best course of action may be to complete the processes of decomposition by exposing the effluent to the action of aerobic bacteria. These are bacteria that thrive in the presence of free oxygen and are capable of breaking down the complex ammoniacal compounds of the sewage into nitrites and nitrates.

This is achieved by distributing the effluent over a bed of hard coke, clinker or stones. The filter bed should be about 4 ft deep and there should be one cubic foot of filtering material for every 40 gallons of estimated daily flow.

Large installations, serving several houses, will provide for the distribution of the effluent over the filter bed by means of perforated rotating arms operated by a kind of water wheel turned by the weight of the effluent flowing through the system. Smaller, domestic systems will have a filter bed that is rectangular in plan covered with a per-

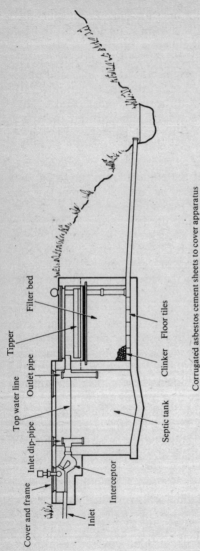

Cover and frame
Inlet
Interceptor
Inlet dip-pipe
Top water line
Outlet pipe
Tipper
Filter bed
Septic tank
Clinker   Floor tiles

Corrugated asbestos cement sheets to cover apparatus

Angle iron
75 mm x 25 mm (3 in x 1 in) battens
Tipper
Filter bed

Fig. 58   Septic tank with filter

forated sheet of corrugated asbestos. A tipper device discharges the effluent first on to one side of the asbestos sheet and then on to the other.

It must be stressed that the purpose of the *filter bed* is *aeration* – not filtration. *Submerged filters* in which the inlet and the outlet of the filter bed are at the same level are totally useless.

After passing through the filter bed the effluent usually passes through a small catch-pit designed to retain the black flecks of 'humus' that it contains, before discharge into a ditch or stream.

Subsoil irrigation

In many parts of the country subsoil irrigation is preferred to filtration as a means of disposing finally of the septic tank's effluent. This method of disposal is most likely to be effective where there is a substantial area of land available and where the subsoil is dry, light and absorbent.

The effluent is allowed to soak into the subsoil via a system of land drain pipes fanning out from the septic tank's outlet. The drains should be laid dead level and will be most effective if they are laid as close to the surface as possible, bearing in mind the depth of the septic tank's outlet.

Land drain trenches should be 50 cm (just over 18 in) wide and should be dug to a depth 30 cm (1 ft) deeper than the level at which the drains are to be laid. They should then be filled with pebbles or clinker to the required level.

It used to be the practice to use unglazed agricultural drain pipes, butt-jointed for the drainage system. Modern perforated uPVC or pitch fibre land drain pipes are easier to lay and much less likely to become clogged with silt.

Having laid the pipes, fill in the trench to a level about 30 cm (1 ft) above the drain pipes with more pebbles or clinker. If this pebble or clinker bed is covered with a sheet of polythene before the top soil is replaced it will reduce the risk of silt washing between the pebbles to clog them.

It is usual for the effluent to discharge directly into the

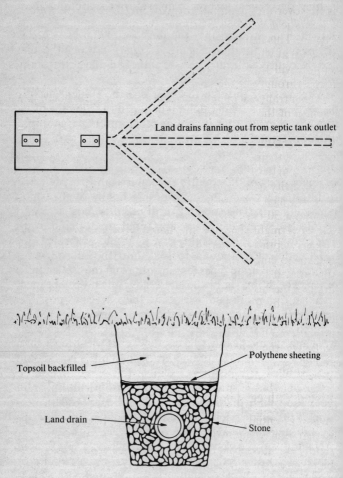

Land drains fanning out from septic tank outlet

Topsoil backfilled

Polythene sheeting

Land drain

Stone

Fig. 59   Land drainage trenches

subsoil drainage system from the septic tank outlet. There is however an advantage in providing a *catch pit* with an automatic *dosing siphon* between the outlet and the drainage system. This catch pit will fill with effluent until it reaches the level at which the dosing siphon comes into action. The siphon will then discharge the effluent, with considerable force, throughout the land drains.

This arrangement prevents the subsoil in the immediate vicinity of the septic tank outlet from becoming sour and 'sewage sick'. It also allows for a 'recovery period' during which the effluent can soak into the soil and the soil bacteria carry out their function of decomposition and purification.

The length of the subsoil drainage system will depend on the porosity of the soil. There is a straightforward practical way in which this can be estimated. Dig a hole 60 cm (2 ft) deep and 30 cm (1 ft) square in plan. Fill with water and leave overnight. The next morning fill the hole with a 25 cm (10 in) depth of water. Measure in seconds the time that it takes to empty. Divide this figure by 250 to establish the time that it takes 1 mm to disperse. Multiply the quotient first by the number of persons that the septic tank serves and then by 0.5. This will give the area in square metres into which the pipes should drain. If the land drain trenches are 50 cm (1 ft 6 in) wide, multiplying this figure by two will give the length in metres of the effluent drainage system.

Living with cesspool or septic tank drainage

A first essential of both cesspool and septic tank drainage is that rain water from roofs and yard surfaces must be rigidly excluded from the system. Surface water drains must be taken either direct to a stream or ditch or to a separate soakaway.

If you have a septic tank remember that septic action is bacterial action. When cleaning gullies, lavatory pans and inspection chambers be very sparing in the use of chemical disinfectants. These are quite incapable of distinguishing between harmful and beneficial bacteria.

The brine wash from a mains water softener should not be permitted to discharge into a septic tank system. Common salt is a disinfectant.

Do not over-use synthetic detergents. These may emulsify the fats in the sewage to produce a soup-like effluent that will clog the filter or the land drain system.

Finally, as a safety precaution, make sure that the covers of the manholes giving access to the cesspool or septic tank are either 'heavy duty' or are securely bolted down. An easily removed cover could prove to be a death trap to an inquisitive child.

Septic tanks need to be de-sludged from time to time. De-sludging twice a year has been recommended but many septic tanks work satisfactorily for several years without attention. This is a job best done by the Local Authority's – or by a private contractor's – cesspool emptier.

Subsoil drainage

The need to lower the level of the subsoil water to prevent the garden from becoming waterlogged and the house damp and cold is not an exclusively rural problem. However it is most likely to be acute where, as is often the case, rural homes are built beside a road running through a valley.

Cultivated fields or pasture land will lie behind these properties and, particularly during periods of heavy and prolonged rain, run-off from this land may turn the gardens into a swamp beyond all possibility of cultivation.

The remedy is to provide a land drainage system that will intercept subsoil water flowing from the fields and divert it harmlessly into the stream or ditch that normally lies between the front of the properties and the highway.

Dig a trench about 75 cm (2 ft 6 in) deep and 50 cm (1 ft 6 in) wide along the rear boundary of the property and a similar trench at right angles to it leading down to the ditch or stream. Fill in the first 30 cm (1 ft) depth of these trenches with pebbles or clinker and lay perforated pitch fibre or uPVC drain pipes upon the bed. The pipes laid along the rear boundary of the property should have a *very slight* fall towards those leading to the ditch. The latter will follow the lie of the land, keeping the same depth of soil above them throughout their length.

Cover the pipes with more clinker or pebbles to within a spade's depth of the surface and then replace the surface soil. There is some advantage in providing an inspection chamber at the junction of the two lines of pipe. This can

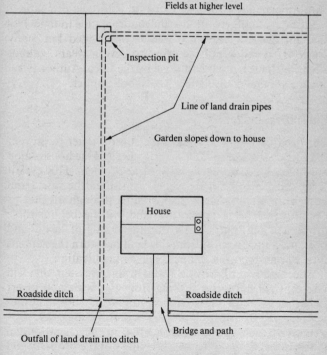

Fig. 60 Subsoil drainage

consist of a simple brick-lined pit with a concrete slab cover

The outlet into the ditch should have a concrete surround and the pipe end should be fitted with a fine mesh grid to prevent rats from entering the land drainage system.

Hedges and ditches

Ditches are, of course, the oldest and most basic form of land drainage. Usually they are flanked by a hedge and disputes can arise as to the exact boundaries of the properties on either side.

This can be important when the ditch needs to be 'brushed out' or when it is proposed to use it for the discharge of septic tank effluent or for the outlet of a land drainage system. In the absence of documentary evidence to the contrary it can usually be assumed that the ditch – not the hedge – marks the boundary.

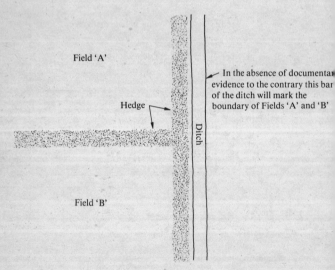

Fig. 61   Hedges and ditches

In the first instance the farmer or land owner will have dug the ditch along the boundary of his property, throwing the excavated soil back on to his own land. He will then have planted a hedge along this artificially produced bank.

While this may have seemed a very sensible arrangement in the first instance it produces problems later. The ditch is *apparently* part of the adjoining field and, except by breaking through the hedge, the person responsible for it has no ready access for inspection or cleansing.

11

Some Plumbing Techniques

The earlier chapters of this book have dealt largely with plumbing design and with routine work of protection and maintenance that should be within the scope of any householder, however limited his experience or his tool kit. Before proceeding to works of replacement or installation it is necessary to master the skills of cutting, bending and joining the pipes used in modern plumbing work and the ways in which they are connected to taps, ball valves and other plumbing equipment.

It must be stressed that the relevant earlier chapters of this book should be studied before work of this kind is undertaken. To attempt alterations or additions to a plumbing system without a thorough understanding of the principles of its design is to invite disaster.

Copper tubing

Light gauge copper tubing is most commonly used for modern hot and cold water supply and central heating installations. It is the development of this tubing, more than anything else, that has brought domestic plumbing within the scope of the d-i-y enthusiast.

Half hard temper tubing is used for above-ground domestic hot and cold water supply. It is sold in straight lengths 6 metres long though most suppliers will be prepared to cut to the lengths required for transport and installation.

Dead soft temper tubing is sold in 20 metre coils. It is used for underground water supply pipes and for microbore central heating installation. The long lengths in which dead soft temper tube is available reduce the number of joints that will be needed. The flexibility of the tubing permits it to be threaded easily through floor and ceiling joists for radiator connections.

Copper tubes are sized by their diameters – nowadays in millimetres. Sizes generally available, together with their Imperial equivalents, are indicated below:

Metric size	Imperial equivalent
12 mm	$\frac{3}{8}$ in
15 mm	$\frac{1}{2}$ in
22 mm	$\frac{3}{4}$ in
28 mm	1 in
35 mm	$1\frac{1}{4}$ in
42 mm	$1\frac{1}{2}$ in
54 mm	2 in

This discrepancy between the metric and Imperial sizes of pipes was referred to in the introduction to this book. It is largely accounted for by the fact that whereas the Imperial measurement was of the internal diameter of the tube the new metric measurement is of the *external* diameter.

The tube sizes most likely to be required for domestic hot and cold water supply are 15 mm, 22 mm and 28 mm.

Non-manipulative (Type A) compression joints

Non-manipulative, or *Type A*, compression joints and fittings provide the easiest – though not the cheapest – way of joining lengths of copper tubing and connecting them to other fittings.

Type A fittings may be made of brass or gunmetal. A coupling for joining two lengths of tube of the same diameter consists of a joint body with, at each end, a soft copper ring or olive retained by a screw-on cap-nut. These are not usually interchangeable between different manufacturers' brands.

To join two pieces of copper tubing, each must first be cut to the required length. This can be done with a hacksaw, using a file to square the end off after cutting. The use of a wheel tube cutter is recommended where a considerable amount of jointing is to be done. This saves time by ensuring a square-ended joint at the first attempt.

Remove all internal or external 'burr' with a file or reamer.

Unscrew the cap nut from one end of the compression coupling and slip it over one of the tube ends, following it with the copper olive. Make sure that the olive is the same way round on the tube as it was in the coupling. With some makes of compression joint this is all-important.

Smear the tube end and the olive with *boss white* or some similar jointing compound and slip into the body of the coupling as far as the tube stop. Screw on the cap nut hand tight.

Repeat this process with the other length of tube and the other end of the coupling.

Using a spanner of the appropriate size on each cap nut, tighten them to compress the olive against the walls of the tube and thus ensure a watertight joint. One complete turn plus one quarter turn of the cap nut is likely to be necessary with 15 mm tube. One complete turn should be all that is needed with larger sizes. Overtightening can lead to leaks.

When first making compression joints it is a good idea to mark the cap nut and the wall of the tube with a piece of chalk before using the spanner. This will allow you to check that you have turned the nut by the right amount. After a little practice you will be able to dispense with these chalk marks.

Extensions to and adaptations of existing water services

are likely to involve the connections of new metric tubing to existing Imperial sized pipes. 12 mm, 15 mm, 28 mm and 54 mm compression fittings are interchangeable with their Imperial equivalents and can be used without adaptation with $\frac{3}{8}$ in, $\frac{1}{2}$ in, 1 in and 2 in Imperial tubing. Where 22 mm, 35 mm or 42 mm tubing is to be connected to $\frac{3}{4}$ in, $1\frac{1}{4}$ in or $1\frac{1}{2}$ in Imperial tubing the new metric compression fittings will need an adaptor to ensure a watertight fit. These adaptors can be obtained from any supplier of compression joints and fittings.

The coupling is, of course, only the most basic kind of compression fitting. A browse through the stock of any

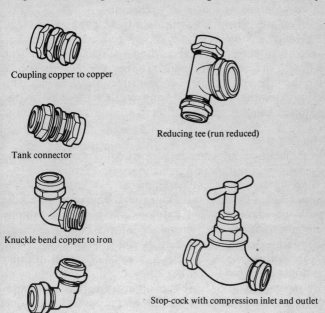

Coupling copper to copper

Tank connector

Reducing tee (run reduced)

Knuckle bend copper to iron

Knuckle bend copper to copper

Stop-cock with compression inlet and outlet

Fig. 62   A few examples from the *Typay* range of non-manipulative compression fittings

builders merchant or d-i-y supplier – or through illustrated catalogues – will reveal reducing couplings, for connecting pipe of unequal diameter, tees, for taking branches from existing pipework, fittings with a compression joint at one end and a swivel tap or ball valve connector (a cap and lining joint) at the other, fittings for tank or cylinder connection, stop-cocks and gate valves with compression joint inlets and outlets, and a wide variety of compression bends to enable changes to be made in the direction of runs of pipe.

There are a number of manufacturers of compression fittings. Their products are sold under various trade names – *Prestex*, *Conex*, *Kontite* and *Typay* for instance. These can all be regarded as thoroughly reliable.

In an area where dezincification (see Chapter 6) presents a problem gunmetal fittings should be used. The inexperienced amateur might find it wise to purchase fittings with a design that makes it possible to see *from the outside* how far the tube end will have to protrude into the joint. This eleminates the guesswork from cutting the pipe to the required length.

Acorn fittings

Manufacturers are constantly researching ways of making even simpler and more reliable joints and fittings.

An interesting recent development has been the *Acorn* joint (Bartol Plastics Ltd, Doncaster). This is an all-plastic push-fit joint made of polybutylene, a material that is totally immune to dezincification and other forms of corrosion and can withstand high temperatures and pressures.

The *Acorn* joint resembles a *Type A* compression joint in appearance *but it does not have to be dismantled to make a watertight joint*. Beneath the cap nut is a *grab ring* of non-corrosive steel that grips the pipe and prevents a blow-out. An 'O' ring seal ensures a watertight joint and the grab ring and 'O' ring are separated by a *spacer washer*.

To make an *Acorn* joint, the pipe end must be cut to length and cleaned as with a *Type A* coupling. All that you do now

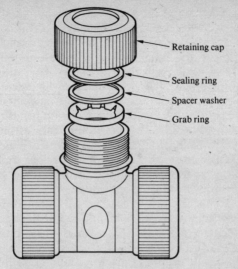

Fig. 63   An *Acorn* push-fit pipe connector

is apply a smear of *Bartol* silicone lubricant to the pipe
end and the 'O' ring and thrust the pipe end home, past
the 'O' ring seal, into the joint. The connection is made!

At the time of writing there is a full range of *Acorn* fittings
available for use with 15 mm and 22 mm copper tube only,
but the manufacturers are planning to develop the fittings
for use with other sizes and other pipe materials.

Manipulative (Type B) compression fittings

Manipulative, or *Type B*, compression fittings are less frequently used in domestic plumbing than *Type A*. They are
however required by Water Authorities where compression
joints are to be made underground. This is because *Type B*
fittings cannot be forced open by internal pressure.

With a *Type B* compression fitting the pipe end itself

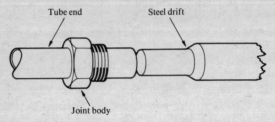

(a) Slip joint body (or cap nut) over tube end and hammer in steel drift to bell out end

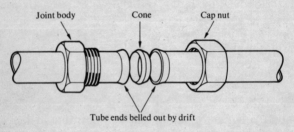

(b) Repeat with other tube end and push ends over cone. Smear cone and tube ends with jointing compound

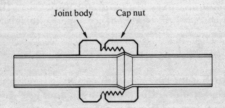

(c) Screw cap nut onto joint body to complete joint

Fig. 64   Making a manipulative compression joint

forms part of the joint, taking the place of the copper olive used with *Type A* fittings.

To make a *Type B* joint the cap nut must be removed and slipped over the pipe end. This pipe end is then 'manipulated'. This is generally done be belling it out with a steel drift, driven in with a hammer. An alternative – used with the well-known *Kingley* joint – is to make a *swage* round the pipe end. A purpose-made *swaging tool* is inserted into the tube end and is turned in a full circle. This action forces a hard steel ball to make a groove round the inside of the pipe and an equivalent ridge or swage round the outside.

When the tube end has been 'manipulated' it must be smeared with joining compound and thrust into – or on to – the body of the fitting. Screwing on and tightening the cap-nut completes the joint.

A *Type B* joint, once made, can be dismantled only by unscrewing the cap nut and then cutting the tube to allow this nut to be pulled forward off its end.

Soldered capillary joints

Soldered capillary joints provide an alternative means of joining copper plumbing work. They are cheaper, and neater in appearance, than compression joints and are no more difficult to make. A blow lamp is however an essential tool. Modern gas-operated blow lamps are quite adequate.

Capillary joints depend for their effectiveness upon the fact that any liquid – in this case molten solder – will be drawn by *capillary action* to fill any very narrow space between two smooth surfaces. The joint therefore consists of a metal sleeve fitting closely over the tube end. *Integral ring* capillary joints incorporate sufficient solder within the fitting itself to make the joint. With the cheaper *end feed* fittings the solder has to be introduced into the joint from a length of solder wire.

Meticulous care and cleanliness are essential to the successful making of this kind of joint.

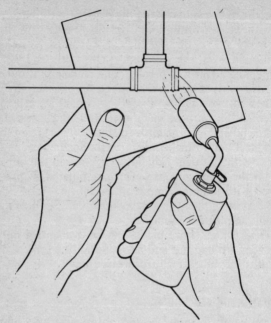

Apply blow lamp flame. Note sheet of asbestos behind fitting

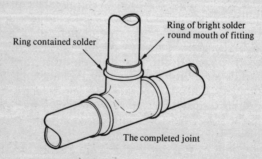

Ring contained solder

Ring of bright solder
round mouth of fitting

The completed joint

Fig. 65   Making an integral ring soldered capillary joint

Cut the tube end square and remove all internal and external burr. Clean the tube end and the interior of the capillary joint with wire wool or with fine abrasive paper. Smear an approved flux over the surface of the tube end and the interior of the fitting. Insert the tube into the fitting to the tube stop and secure in position.

With the integral ring fitting all that now needs to be done is to apply the flame of a blow lamp to the joint. The solder will melt and flow to fill the narrow space between the tube end and the fitting. The joint is made when a silver ring of solder can be seen *all round* the mouth of the fitting.

With an end-feed fitting, heat the joint with the flame of the blow lamp until flux vapours can be seen escaping. This indicates that the fitting and tube end are hot enough for the solder to run. Apply soft solder wire (50% tin and 50% lead) to the mouth of the fitting. It will melt and will be drawn in to make the joint. Once again, the appearance of a ring of solder all round the mouth of the fitting indicates that the joint is made.

It can be helpful to know roughly how much solder is likely to be needed when making an end-feed joint. A 15 mm joint needs about $\frac{1}{2}$ in of solder wire, a 22 mm joint $\frac{3}{4}$ in and a 28 mm joint 1 in. Bend this length of wire over before you begin work and you will know that when the whole of the bent-over piece of wire has been melted and drawn into the joint it is nearing completion.

Once the joint has been made leave it undisturbed until it is cool enough to touch.

There are two or three points that must be noted in connection with the use of soldered capillary joints and fittings.

Never forget the potential fire hazard involved in the use of a blow lamp. Put a sheet of asbestos or a piece of fibre-glass mat between the fitting and the skirting board or any other flammable surface behind it. Keep the flame well away from any plastic equipment – particularly the acrylic plastic which modern baths and wash basins are sometimes made of.

Where more than one joint has to be made to one fitting – the two ends of a coupling or the three ends of a tee junction for instance – it is best to make all joints at the same time. If this cannot be arranged a damp cloth should be bound round any joints already made when making a later one. This will prevent the solder from re-melting.

Metric capillary fittings *cannot* be used with Imperial tube. Adaptors are available.

Other methods of jointing

Other methods of jointing copper tubing include silver soldering, brazing, bronze welding and copper welding. These all involve the use of oxy-acetylene equipment capable of producing very high temperatures. They are rarely, if ever, used in domestic plumbing.

There are, on the market, tube expanders that can be inserted into the end of a length of copper tube and used to increase the tube's diameter sufficiently to permit the end of another tube to be inserted. The expanded tube end becomes, in effect, a built-in end-feed soldered capillary joint. This technique can be well worthwhile for anyone faced with a major plumbing operation.

Bending copper tubing

Knuckle and *easy* bends of various kinds are to be found in every illustrated catalogue of compression and soldered capillary joints and fittings. It is however possible to make easy bends in 15 mm and 22 mm copper tubing by hand. This saves money and results in a neater and more professional looking job.

If you take a length of copper tubing and bend it over the knee you will find that, although the bend is made easily enough, the walls of the tube collapse and the tube becomes elliptical in section at the point at which it is bent.

This collapse can be prevented by providing support for the tube walls as the bend is made. Bending machines provide *external* support. Bending springs support the walls internally.

Bending springs are made of hardened steel and have a loop at one end. The spring – of the appropriate size of course – must be greased before being inserted into the tube to the point at which the bend is to be made.

Then, with the spring in place, bend the tube over the knee. Best results will be obtained by overbending to a few degrees and then bringing back to the required angle. To remove the spring, insert a bar into the loop at the end, twist to reduce the spring's diameter, and pull.

Bending in this way may leave a few minor kinks in the inner surface of the bend. These can be 'dressed' out afterwards with a hammer. On no account attempt this before you have withdrawn the spring. If you do you may find the spring to be immoveably stuck within the tube!

In theory tubes up to 28 mm in diameter can be bent with the use of an appropriate spring. The beginner would however be wise to limit himself to easy bends in 15 mm or 22 mm tube.

Where difficult 'out of sight' bends need to be made – on the cold water supply pipes to a bath or basin for instance – *U-can copperbend* can be extremely useful. This is, in effect, a length of corrugated copper tubing that can be bent by hand without recourse either to a bending machine or a spring. It is obtainable in 22 mm and 15 mm sizes and may have either two plain ends, for connection to compression fittings, or one plain end and one end provided with swivel tap connector for connection either to a tap or a ball valve.

Pipe fixing

Lengths of copper tubing must be secured to walls or skirtings with suitable pipe clips. Never skimp pipe support. To do so will be to invite noise and vibration.

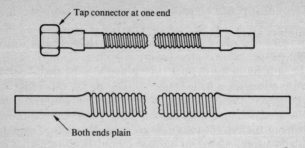

Tap connector at one end

Both ends plain

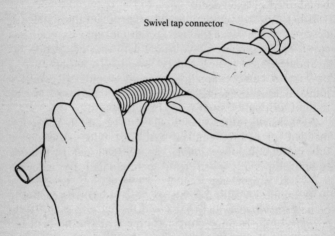

Swivel tap connector

Fig. 66   U-can copperbend

Recommended distances between pipe clips are as follows:
  15 mm copper tube – clips at 1.2 metre centres on horizontal runs and 1.5 metre centres on vertical ones.
  22 mm and 25 mm copper tube – clips at 2 metre centres on horizontal runs and 2.5 metre centres on vertical ones.

Stainless steel tubing

Light-gauge stainless steel tubing is obtainable in the same sizes as light gauge copper tubing and can be used for all the purposes for which copper tubing might otherwise be used. It has the advantage that it can be connected to galvanised steel or copper plumbing equipment without involving the risk of electrolytic corrosion. It is thus particularly useful where an old galvanised steel system is to be adapted or extended.

It has not, in my opinion, enjoyed the popularity that it deserves. It has an attractive appearance and its price has consistently compared favourably with that of copper.

Stainless steel tubing can be joined with either *Type A* or *Type B* compression joints or with soldered capillary joints. The *Acorn* joints referred to earlier cannot, at the time of writing, be used with these tubes.

A hacksaw, rather than a wheel tube cutter, is recommended for cutting stainless steel tube. This is because a tube cutter could *work harden* the tube end making it liable to split, especially when being 'manipulated' for use with a *Type B* compression joint.

As stainless steel is a harder material than copper it may be necessary to exert a little more force when tightening up compression cap nuts to ensure a watertight joint.

When making soldered capillary joints with stainless steel tube you should use a flux based on phosphoric acid, not one with the usual chloride base. This can burn the fingers so always apply it with a brush or spatula. It too can be difficult to obtain. Don't purchase capillary fittings for use with stainless steel until you have made sure that you can also buy the necessary flux.

In making the joint the blow lamp flame should be directed solely on the fitting – never on the stainless steel tube.

A – possibly limited – range of chromium plated compression and capillary fittings is available for use with this kind of tubing.

Stainless steel tubing is more difficult to bend than copper. 22 mm and 15 mm tube can be brought to easy bends in a bending machine. 15 mm tube *can* be bent with the use of a bending spring but most installers prefer to stick to straight lengths, making any necessary changes of direction with appropriate compression fittings.

Screwed iron pipes

Heavy, wrought iron or malleable cast iron pipes, usually galvanised as a protection against corrosion, were frequently used in pre-war plumbing work. They were joined by screwed joints and, when a length of pipe had to be cut, it was necessary to use stocks and dies to make a male thread on the cut end.

These pipes cannot be bent and are clumsy and unsightly in appearance. They are never used nowadays in new plumbing work but there are, of course, many thousands of iron plumbing systems still in existence.

Provided that they are not already corroded, systems of this kind can be extended by the use of stainless steel – but not copper – pipes. Compression fittings are available with a screwed thread at one end and a compression joint, for connection to the new stainless steel tube, at the other.

To make a screwed joint watertight, bind one or more turns of *PTFE* plastic thread sealing tape round the male thread before screwing it home. PTFE tape is obtainable from all builders merchants in rolls resembling a roll of surgical tape.

Lead pipe

Lead is the traditional plumber's material and has been used for over 2000 years.

However, because of its high price and its potential health

hazard lead is never used today in new plumbing installations. This is wholly to the advantage of the d-i-y plumber. The malleability of lead makes it easy to use by the experienced worker but prevents it from being joined by mechanical means. Unskilled householders are unlikely to acquire the necessary experience to make the *wiped soldered joints* used with lead pipe neatly and effectively.

Lead was traditionally used for plumbing because of its resistance to corrosion. In contact with the atmosphere a film of lead oxide quickly forms on exposed surfaces which protects the metal from further oxidation. Consequently there are still many thousands of houses with lead hot and cold water systems, so the professional plumber still needs skill in handling this metal for repair, replacement and extension work.

Sizes of lead pipe have remained virtually unchanged by metrication. Lead pipe, unlike copper and stainless steel, is designated by its *internal* diameter. Metric sizes, set out below, are more or less straight translations of the old Imperial sizes.

Imperial size	Metric size
$\frac{3}{8}$ in	10 mm
$\frac{1}{2}$ in	12 mm
$\frac{3}{4}$ in	20 mm
1 in	25 mm
$1\frac{1}{4}$ in	32 mm
$1\frac{1}{2}$ in	40 mm
2 in	50 mm

Wiped soldered joints

Lengths of lead pipe are joined by wiped soldered joints. It is easy enough to describe the way in which joints of this kind are made but the novice is unlikely to make a neat, watertight joint.

Careful preparation of the pipe ends is the prerequisite

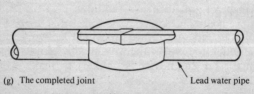

(a) Opening the end — Tanpin

(b) Rasping away the arris

(c) Shaving the socket — Shavehook — Tarnish or smudge applied

(d) Chamfering the spigot end

(e) Shaving the spigot

(f) Joint ready for wiping — Smeared with tallow

(g) The completed joint — Lead water pipe

Fig. 67 Making a wiped soldered joint

of success. This is how it is done. Cut the ends squarely and
remove all burr with a rasp. Then form a socket in one end
using a *tanpin* or hardwood cone. Rasp away the external
edge – sometimes called the *arris* – of this opened-out end.
Form the other pipe end into a *spigot* that will fit closely
into this socket by rasping away its outer edge to the angle
of the tanpin.

Coat about three inches of each pipe end with *plumbers
black* or tarnish to which solder will not adhere. Mark the
extent of the joint on each end. A wiped joint between 12 mm
($\frac{1}{2}$ in) or 20 mm ($\frac{3}{4}$ in) pipes should have a total length of
70 mm. A joint between 25 mm (1 in) pipes should have
a length of 75 mm.

Shave away the tarnish and the surface coating of lead
oxide from the area of the joint to leave the bright metal
underneath exposed. Smear this *immediately* with tallow to
prevent further oxidation. Push the spigot of the one pipe
into the socket formed in the end of the other and secure
firmly for wiping.

For this you will need a blow lamp, a *moleskin* wiping
cloth and a stick of *plumbers solder*. This contains roughly
two parts of lead to one of tin and has a higher melting
point (about 230°C) than the solder used with capillary
soldered joints.

Apply the flame of the blow lamp to the lead pipe on
each side of the joint, moving it to and fro until the whole
of the joint area has been raised to a temperature above
that of the melting point of the solder. 'Tin' the shaved
surface by rubbing the solder stick lightly over it. Some
solder will run, by capillary attraction, into the space
between the spigot and socket of the joint.

Apply more heat. The solder stick will soften, and blobs
of solder will be released. These must be moulded, while
still plastic, with the wiping cloth to a neat 'wiped joint'
appearance.

Connecting lead to copper tubing

The d-i-y plumber is more likely to need to connect a new length of copper tubing to an existing lead pipe. He will have to do this if he wishes to remove an old lead system and replace it with a copper one, or if he wishes to extend an existing length of lead pipe.

The connection can be more difficult than might appear at first sight.

All manufacturers of compression and capillary joints and fittings include a lead-to-copper connector in their range. However, the lead end of this joint is simply a plain length of brass or gunmetal tube that must be connected to the lead pipe with some kind of soldered joint.

Wiping a joint on to a lead-to-copper connector

The way in which professional plumbers make the connection is by means of a wiped soldered joint similar to the one already described.

The lead pipe end is opened up with a tanpin, rasped, tarnished and cleaned to form a socket for the brass connector.

The next task is to prepare the fitting for connection. Score its 'tail' with a medium cut file to remove impurities. Lightly smear the scored end with tallow and apply tarnish to the other end to mark the limit of the joint.

'Tin' the scored end of the joint by covering with a coat of general purpose solder. This should have equal parts of lead and tin. Use a large copper bit to apply the solder.

Place the tinned end in the socket that has been formed in the lead pipe end, secure in position with wooden splints thrust through the fitting and into the lead pipe, and make the 'wiped joint'.

This can be a difficult job, particularly when the connection between the lead and copper pipes is vertical. When extending a lead system with copper tubing many competent

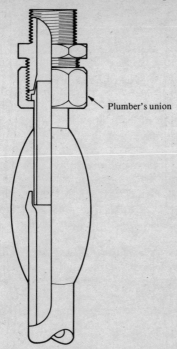

Fig. 68   A wiped lead-to-copper or lead-to-iron connection

amateur plumbers prefer to have professional help with the actual lead-to-copper connection, before carrying on with the rest of the work single-handed.

The Staern or soldered spigot joint
The *Staern* or *soldered spigot joint* provides a means of making a vertical lead and copper joint that is, in my opinion, within the capacity of most d-i-y enthusiasts.

Unfortunately, because it is rarely used nowadays by professional plumbers, the amateur may have difficulty in

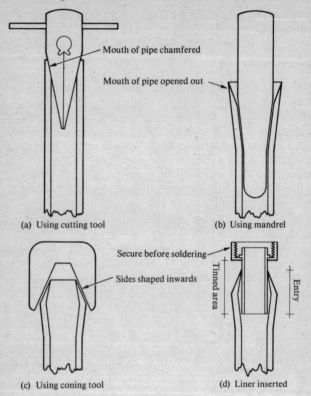

(a) Using cutting tool

Mouth of pipe chamfered

Mouth of pipe opened out

(b) Using mandrel

Secure before soldering

Sides shaped inwards

Tinned area

Entry

(c) Using coning tool

(d) Liner inserted

Fig. 69   Making a *Staern* or soldered spigot joint

getting hold of the necessary tools. These, and the way in which they are used, are illustrated in Figure 69.

Fill the small end of the cutting tool with tallow to prevent shavings falling into the pipe. Then introduce it into the end of the lead pipe and rotate gently to chamfer the edge of the pipe end down to 2 mm or less. Clear the shavings from the tool as the operation proceeds.

Next, smear the *mandrel* with tallow and hammer into the tube end to open it up to a minimum depth of 32 mm. Withdraw the mandrel. Tap a hollow hardwood cone over the chamfered pipe end to shape the sides inwards. Continue this process until the tail of the compression fittings can be pushed into it to give a fairly tight fit. File off the sharp mouth of the socket to give a clean, square edge.

The tail of the compression fitting must now be prepared. Rasp and tin its tail and smear with *flux paste* before inserting it into the socket.

Apply the flame of a blow lamp to the joint, concentrating the heat on the tail of the compression fitting, just above the mouth of the socket. Apply solder wire to the heated area. This will melt and will be drawn, by capillary attraction, into the confined space between the spigot and socket. The joint is complete when a bright ring of solder can be seen all round the mouth of the socket and when no more bubbles of flux appear.

The cup and cone joint

The cup and cone joint is easily made but is unsuitable for pipes carrying water under pressure. It may however be used with gas pipes or with waste pipes.

Drive a hardwood cone or turn-pin into the end of the lead pipe until the tail of the compression fitting can be

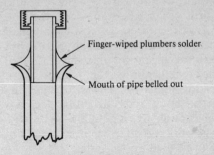

Finger-wiped plumbers solder

Mouth of pipe belled out

Fig. 70   A finger-wiped joint

inserted into it to a depth equal to about half its diameter.
Prepare the end of the tail by rasping, tinning and applying
flux, then insert it into the belled out socket and fix securely.

Run fine solder into the cup to fill it.

A *Taft* or *finger-wiped joint* is made in the same way except
that *plumber's wiping solder* is used and the solder is wiped,
while still plastic, up the tail of the compression fitting. *This
joint is unsuitable for pipes carrying water under pressure.*

Plastic materials

A wide variety of plastics are used in modern plumbing for
water supply and waste pipes, above and underground
drains, roof drainage and for wash basins, baths, ball valves,
stop-cocks, taps, waste traps and cold water storage cisterns.

Pipes and fittings made of these materials are light in
weight, easily cut and fitted, smooth and self-cleansing and,
of course, totally non-corrodible.

They have also be consistently cheaper in price than their
metal equivalents. However, plastics are oil based and their
price has been – and will continue to be – affected by the
rising price of oil.

Unplasticised polyvinyl chloride
Unplasticised polyvinyl chloride (usually abbreviated to
uPVC, PVC or, simply, vinyl) is widely used in domestic
plumbing. It is perhaps best known as the material which
waste and drainage systems are commonly made of, but it
can also be used for cold water supply and distribution pipes.

There are three means by which lengths of uPVC tubing
can be joined together or to other plumbing fittings – push-
fit compression joints, solvent weld joints and ring-seal
joints. For the supply and distribution of cold water under
pressure solvent weld joints must always be used. For waste
and drainage work it is quite common for all three methods
of jointing to be used in one system.

Push-fit compression joints

This is the simplest form of connection. It is used for connecting waste pipes to traps, for connecting short lengths of waste pipe and for overflow or warning pipes.

The joint resembles the *Type A* (non-manipulative) compression joints used with copper tubing except that it is made of plastic and a rubber or plastic sealing ring replaces the soft copper olive. To make such a joint the pipe end must first be cut square. The cap nut of the fitting is then loosened – not completely unscrewed – and the pipe end thrust into it to the pipe stop. Tightening up the cap nut completes the joint.

Solvent weld joints

Solvent weld jointing resembles soldered capillary jointing of copper tubing in some respects but no blow lamp or other source of heat is required.

The pipe end must be cut square with a hacksaw or other fine toothed saw and all swarf or burr removed with a file. Insert the pipe end into the solvent weld socket as far as the pipe stop, and mark the depth to which it is inserted with a pencil.

Withdraw the pipe and roughen its outer surface as far as your pencilled mark with medium grade abrasive paper. Roughen the inside surface of the socket in the same way. Do *not* use steel wool for this job. It will polish the plastic surfaces instead of roughening them.

Clean the pipe end and the interior of the socket with a spirit cleaner and degreaser approved by the manufacturer of the solvent weld joint. Wipe off with a clean tissue.

Next apply an even coat of solvent cement to the pipe end and to the interior of the socket. Use a wooden spatula to stroke the cement *along* (rather than round) the surfaces to be joined together.

Thrust the pipe end into the socket with a slight twist. Hold in position for a few seconds. Although the joint can

1   Cut the pipe straight and square using a hacksaw
2   Clean off swarf and burr inside and out using a half round file
3   Assemble the fittings then check for length and alignment. Mark pencil lines to avoid
    errors

4   Roughen the external surface of the pipe end and internal surface of the socket with
    medium abrasive paper. Clean socket and pipe ends with approved spirit cleaner and
    degreaser. Wipe clean
5   Using a spatula to distribute solvent around the pipe exterior and socket interior
6   Push the pipe into the socket with twisting action. Hold for 15 seconds
7   Remove surplus cement immediately with dry, clean cloth after each joint is made

Fig. 71   Making a solvent weld joint with Marley pvc tube

be handled safely after two or three minutes it should not
be put into use for at least twenty-four hours.

When buying solvent weld fittings, do read the instructions provided with them. They could vary slightly from
these. One maker, for instance, advises that the pipe end
should be chamfered with a file to an angle of about 15°.
Another says that it is better *not* to twist the pipe as it is
thrust into the socket.

When fitting long lengths of uPVC waste pipe, provision
must be made for thermal movement as warm wastes flow
through them. A 39°C increase in temperature will result
in a 13 foot length of uPVC waste pipe expanding by $\frac{1}{2}$ in.

To accommodate this movement, expansion couplings
should be provided at 6 ft intervals in long lengths of waste

pipe. Such a coupling has an ordinary solvent weld joint at one end and a ring seal joint at the other. Before thrusting the pipe end into the ring seal joint make an insertion mark with a pencil about two inches from its end. Inserting the pipe to this mark will ensure that sufficient room is left for thermal expansion.

Ring seal joints

Apart from their use in expansion couplings, ring seal joints are used with uPVC pipework mainly for the large diameter pipes used for the main stack pipes of single stack drainage systems and for underground drains.

Cutting the pipe end square – the first step that must be taken with all forms of jointing – can be difficult with large diameter pipes. The best way to make sure of a square cut is to lay a sheet of newspaper over the pipe, draw the edges

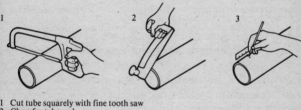

1 Cut tube squarely with fine tooth saw
2 Chamfer tube end
3 Mark depth of tube in socket

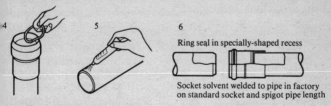

Ring seal in specially-shaped recess

Socket solvent welded to pipe in factory
on standard socket and spigot pipe length

4 Insert ring joint in socket
5 Apply petroleum jelly to tube end
6 Align tube end to socket and push home   (see *Figure* 71)

Fig. 72   Making a ring-seal joint

together underneath it and secure them with a paper clip. The newspaper can then be used as a template for your saw.

Draw a line round the pipe about ⅓ inch from its cut end and chamfer back to this line with a rasp or similar shaping tool. Insert the pipe into the socket to the pipe stop and mark the depth of insertion with a pencil. Withdraw the pipe and make another pencil mark ⅓ inch nearer to the end of the pipe than the original one. When the joint is made the pipe is inserted to this second mark so as to leave ⅓ inch for expansion.

Clean the recess inside the socket of the joint and insert the sealing ring. Smear a small amount of petroleum jelly round the pipe end to lubricate it. Then push the end firmly past the joint ring into the socket. Adjust the pipe so that your second pencil mark coincides with the edge of the socket.

Bending uPVC tubing

Easy bends *can* be made in small bore uPVC tubing by hand after *very gently* heating the area of the tube to be bent with the flame of a blow lamp. This takes practice however and the d-i-y enthusiast would generally be well advised to use purpose-made solvent weld bends when changes of direction need to be made.

Polypropylene tubing

Polypropylene is an alternative to uPVC for the manufacture of traps, rainwater and waste drainage systems. It is more resistant to heat and to chemical action and is therefore frequently used for waster systems from commercial and industrial premises.

Polypropylene cannot be joined by solvent welding. Ring seal or push-fit joints must be used throughout any system made of this material.

Polythene tubing

Polythene is used for the manufacture of cold water storage cisterns, feed and expansion tanks, water supply and distribution and waste drainage pipes. Like uPVC and polypropylene it cannot be used for the conveyance of hot water under pressure.

Polythene tubing, unlike that made of uPVC or polypropylene, is not rigid and is sold in long coils. This has both advantages and disadvantages.

Lengths of polythene tubing sag if they are not given adequate support and, for horizontal runs, continuous support is essential. Failure to provide this will produce an unsightly appearance that is exaggerated by the thickness of the material.

On the other hand the long lengths in which polythene tubing can be bought mean that it can be laid underground without out-of-sight joints which are always potential points of leakage. The thickness and nature of the pipe walls also give polythene tubing a built-in resistance to frost. Should water freeze in a polythene tube the elasticity of the material allows the resultant expansion to take place without damaging the pipe. This makes polythene particularly useful in a garden water supply or – for farmers and camping site operators – for taking a water supply to stand-pipes or cisterns remote from the water source.

The elasticity of polythene tubing also means that it can play a part in silencing noisy water supply systems. Vibration and water hammer in a rising main can often be eliminated by substituting polythene tubing for copper for the final yard or so of rising main before connection to the ball valve supplying the main cold water storage cistern.

Polythene tubing has not, at the time of writing, been metricated. It is still designated by its Imperial internal diameter – $\frac{1}{2}$ in, $\frac{3}{4}$ in, 1 in and so on.

It is joined by *Type A* (non-manipulative) compression

fittings similar to those used with copper and stainless steel tubing. However, because of the relative softness of polythene, a metal insert must be provided to prevent the pipe walls collapsing when the cap-nut is tightened. Polythene tubing can vary in wall thickness so, when purchasing fittings, take a sample of the tube along to make sure that the correct size is obtained.

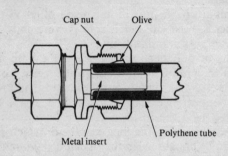

Cut-away section of polythene joint

Fig. 73   *Kontite* compression coupling used with polythene tubing

To make a joint in polythene tubing the tube end must, of course, be cut square. Unscrew the cap-nut of the fitting and slip it, followed by the olive, over the tube end. Next push in the metal insert. If this should prove difficult, warming the tube end will help. Push the tube end into the body of the compression fitting and tighten the cap-nut by hand. Finally give the nut a further one and a half to two turns with a spanner.

Bending polythene tubing

Polythene tubing can be bent cold to easy bends – for a change of direction in an underground supply pipe for instance – but, if left unsupported, will spring back again.

Permanent unsupported bends can be made by heating the tube by playing the flame of a blow lamp *very gently* along the area of the bend or by immersing it, for at least ten minutes, in constantly boiling water.

Pitch fibre pipes

Pitch fibre is not, of course, a plastic material. However, when used for underground drainage, pitch fibre pipes have similar characteristics to the uPVC pipes that may also be used in this situation.

No concrete base is required as the pipes and their joints

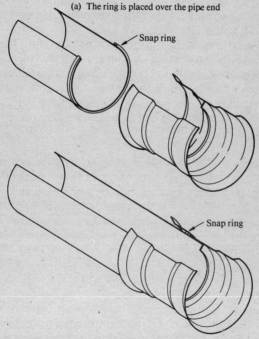

(a) The ring is placed over the pipe end

Snap ring

Snap ring

(b) The coupling is pushed over the snap ring onto the pipe end

Fig. 74 Snap ring connection of pitch fibre pipes

have sufficient resilience to accommodate slight subsoil settlement without damage. A proper base must however be provided and in some cases it will be necessary to bring gravel to the site for this. Care must be taken to avoid damage when infilling.

It used to be the practice to use *fusion jointing* to connect lengths of pitch fibre pipe. The pipe end was hammered into a specially-made, tapered fibre socket. This action fused the exterior of the pipe wall to the interior of the socket.

Nowadays *snap ring* jointing is always used in domestic drainage work.

The pipe end is cut square with an ordinary wood saw. Lubricating the saw blade with water will prevent the clogging that would otherwise take place.

The snap ring is then placed on the pipe end. The ring must have its flat surface in contact with the pipe and it must be placed square to the pipe's axis. Push the coupling over the ring and the pipe end to force the ring to roll along the pipe. This will compress the ring and force it to 'jump' – with a movement that can be felt as it takes place – into its final position.

12

Plumbing Work in the Kitchen

If you have grasped the principles of hot and cold water supply and drainage design outlined in the earlier part of this book and have mastered the techniques set out in the previous chapter you can now carry out straightforward plumbing replacements or improvements in the kitchen and bathroom.

So far as plumbing replacements are concerned it is a common experience of professional plumbers and d-i-y enthusiasts alike that disconnecting and removing the old appliance can often be far more difficult than installing the new one. Where appropriate therefore an attempt is made in the following pages to describe both removal and installation.

Fitting a new sink unit

Fitting a new sink is likely to be an early priority for anyone buying a house which is more than fifteen to twenty years old. The existing kitchen sink is likely to be a glazed ceramic one of *Belfast* pattern. It will have a built-in weir overflow and a detachable wooden drainer. Hot and cold water supply will be provided from bib-taps projecting from the glazed tiles covering the wall behind it.

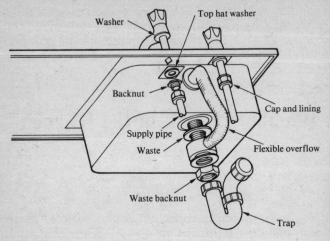

Fig. 75   A modern sink unit

The new householder will undoubtedly wish to replace this with a modern sink unit having a combined sink top and drainer, which is available in enamelled pressed steel or in stainless steel. Stainless steel sink tops are more expensive but they are also more hard wearing and are far less susceptible to accidental damage.

The old ceramic sink must, of course, be removed before the new suit is fitted. However, to avoid an unnecessarily long disruption to the household's water supply the first task should be to fit the taps, waste and overflow to the new sink.

Either individual pillar taps or a mixer may be fitted. Before buying a mixer make sure that it is a *sink* mixer (see Chapter 4) and that the distance between the two tails coincides with the distance between the holes provided for them in the sink top.

To fit, remove the back nuts from the tails of the taps or mixer, slip a flat plastic washer over each tail and thrust them through the holes at the back of the sink. Another washer must be pushed on to each tail, beneath the sink, before the back nuts are screwed on and tightened. As the sink is of thin material the shanks of the taps will protrude through the holes. To accommodate these protruding shanks, *top hat* or *spacer* washers should be used.

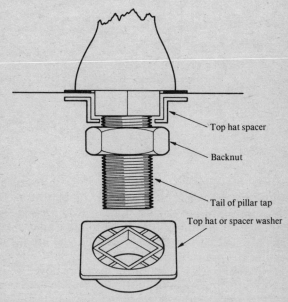

Top hat spacer

Backnut

Tail of pillar tap

Top hat or spacer washer

Fig. 76   Fitting a pillar tap into a stainless steel sink top

As modern sink tops do not have a built in overflow, they should be fitted with a combined waste and overflow fitting. The overflow outlet is connected to the waste, above the level of the trap, by means of a flexible tube.

Smear a non-setting mastic such as *Plumbers Mait* liberally round the waste outlet in the sink and bed the flange of the waste fitting down on to this. Slip a large

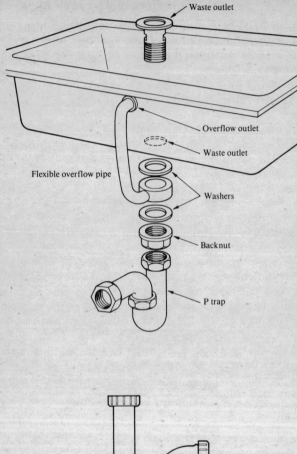

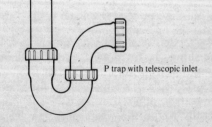

Fig. 77 Sink waste components

plastic washer over the tail of the waste fitting. Screw on the back-nut and tighten securely.

It may be that the trap of the old sink will be suitable for the new one. If not, it would be wise to choose a new trap with a telescopic inlet. This will make it possible for the outlet to be adjusted to the level of the existing waste pipe.

A double sink needs only one trap. This is fitted to the waste outlet of the section of the sink nearer to the waste pipe. The waste outlet from the other section is taken, in a similar way to the overflow pipe, to connect to the first waste fitting above trap level.

Removing the old sink

The next task is to remove the old sink. At this stage there is still no need to cut off the water supply to the taps.

Remove the drainer. Unscrew the large nut securing the waste outlet to the trap and pull the trap to one side. Before the sink can be lifted from its cantilever support it will usually be necessary to break the seal between the back of the sink and the tiled wall behind. This can be done with a cold chisel and a hammer.

You will need help lifting the sink and taking it outside, as it will be very heavy.

The final step is to saw off, flush with the wall, the heavy iron cantilever brackets on which the old sink was supported.

The water supply must now be cut off to remove the old bib-taps and prepare the water supply pipes for connection to the the new unit's taps.

Turn off the main stop-cock. Place a bucket under the cold tap and drain off the few pints of water that will be left in the rising main.

Unless there is a convenient gate valve on the hot water distribution pipe, you will have to drain the cold water storage cistern to cut off water supply to the hot tap. Tie up the ball valve to this cistern so that no more water can enter and open up the bathroom *cold* taps. Provided that you have an indirect cold water system there will be no need to drain away all the stored hot water in the cylinder.

Turn on the bathroom hot taps only when the cold ones have ceased to flow. Finally, with the bucket under it, drain from the kitchen hot tap the few pints of water that will be left in the distribution pipe.

Unscrew and remove the old taps.

The water supply pipes will almost certainly be 'chased' into the wall behind the sink. These must be dug out with cold chisel and hammer and pulled forward to serve the new sink.

Fitting the new sink unit

Measure, very carefully, the vertical distance between the floor and the tails of the taps fitted into the new unit. Measure the same distance from the floor up the hot and cold water supply pipes and cut them off at that level. You may have to cut off another inch or so later to accommodate the compression fittings with their swivel tap connectors but it is better, in the first instance, to have the pipes too long rather than too short.

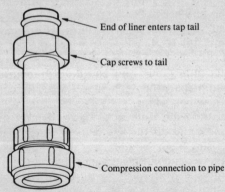

End of liner enters tap tail

Cap screws to tail

Compression connection to pipe

Fig. 78   Swivel tap connector with compression inlet

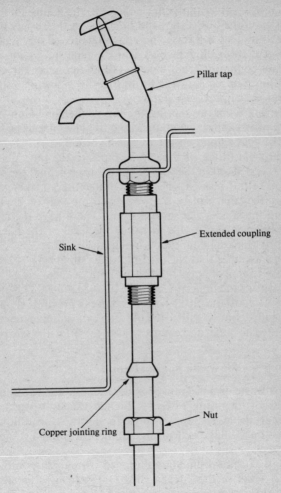

Fig. 79 A *Kontite* tap extension piece

Place the new sink unit in position.

Screw a swivel tap connector (Cap and lining joint) to the tail of one of the taps and hold one of the supply pipes against it. This will reveal how much more – if any – needs to be cut from the pipe end. Cut the pipes to fit, unscrew the tap connector from the tap tail and connect the compression joint ends of the fittings to the pipe ends as described in the previous chapter.

Having done this, spring the tap connector linings into the tap tails and screw up the cap nuts. The liner has a fibre washer that should ensure a watertight joint.

The final fitting of the supply pipes to the tap tails has to be carried out in the very confined space behind the sink. To make this task easier some manufacturers make extension pieces that can be screwed on to the tails of the taps before the sink is put into position. These have the effect of bringing the tap tails to below the level of the base of the sink, making connection much easier.

This assumes that the existing water supply pipes are of $\frac{1}{2}$ in or 15 mm copper. If either or both of them are of lead the swivel tap connector must have a *lead inlet*. This must be soldered to the pipe as described in the previous chapter.

Finally, connect either the old trap or a suitable new one to the waste outlet and connect this to the branch waste pipe.

Fitting a sink waste disposal unit

A sink waste disposal unit provides a simple and convenient means of disposing of 'soft' kitchen wastes – vegetable peelings, tea leaves, apples cores, food scraps, dead flowers and so on for flat dwellers, and all those without a compost heap.

Waste disposal units are plumbed permanently into the waste outlet of the kitchen sink. Operated by a capacitor start 420 watt ($\frac{1}{2}$ hp) induction motor, powerful steel blades grind soft household wastes into a slurry that is then flushed into the drainage system by turning on the sink cold tap.

Although there are models that will fit a standard 38 mm (1½ in) sink outlet, most require an 89 mm (3½ in) outlet. The larger outlet is obviously to be preferred as pushing kitchen wastes through a 38 mm (1½ in) diameter outlet can be a time consuming task that robs the unit of much of its advantage.

Both stainless steel and enamelled pressed steel sinks are available with an 89 mm (3½ in) outlet. The outlets of standard stainless steel sinks can be enlarged to 89 mm with a special cutter usually obtainable on loan or hire from the supplier of the disposal unit. However an attempt to enlarge the outlet of a standard enamelled pressed steel sink is not recommended because of the high risk of chipping the enamel with resultant corrosion.

Plumbing in the unit presents no difficulty. A clamp seal – a rubber or plastic washer of the appropriate size – is slipped over the waste outlet beneath the sink followed by the clamp that supports the unit. This is then fitted to the clamp either with bolts or by a snap fastener.

A standard 'U' or 'S' trap (*not* a bottle trap) must then be connected to the unit's outlet. The waste pipe from this trap may then discharge to a yard gully or – where there is a single stack drainage system – to the main soil and waste stack. If the waste discharges over a gully it is important to ensure that the waste pipe outlet extends to below the gully grid, either through a back or side inlet or through a slot in the grid itself.

The electrical power supply to the unit can be taken from a 13 amp outlet, either through the fused plug and socket or a switched fused connection unit. The latter arrangement is to be preferred.

The most common fault to which sink disposal units are prone is jamming – usually as a result of misuse. Many modern units have a reversible action motor which permits jamming to be cleared by flicking the reversing switch and restarting the motor. Jamming may also result in the motor overheating and operating a thermal cut-out. In this event it will probably take about five minutes for the motor to cool sufficiently to allow it to be re-started.

Models without a reversible action are provided with a key to free the jammed unit. Switch off the electrical connection unit or pull out the plug before using this key. By the time the unit has been freed it is likely that the motor will have cooled sufficiently to be restarted.

Plumbing in an automatic washing machine

The installation of an automatic washing machine is likely to be the modern householder's first plumbing priority after the provision of an up-to-date sink unit. This will need both a hot and cold water supply and means of drainage.

The usual way to provide water supplies is to *tee* into the hot and cold supply pipes serving the kitchen sink and to connect washing machine stop-cocks to the short branch pipes taken from these tees. In describing the way that this is done I shall, once again, assume that the kitchen water supplies are taken through 15 mm ($\frac{1}{2}$ in) copper tubing.

The chances are that these pipes will be both parallel and set close together on the kitchen wall. To ensure minimal interruption of plumbing services tackle one pipe at a time.

Cut off the water supply and drain the pipe as described in fitting a sink unit. Decide the level at which the tee junction is to be fitted and, with a tube cutter or hacksaw, cut out an 18 mm ($\frac{3}{4}$ in) section of pipe at this point. Some water will probably flow out as you make the first cut. Be prepared for it.

Make sure that the pipe ends are squarely cut and file away any internal or external burr.

Take an equal ended 15 mm compression tee. Remove the cap nut and olive from one of the 'run' ends of the tee and slip them over the upper section of the cut pipe. A clothes peg can be clipped over the pipe beneath them to prevent them from slipping down. Repeat this process with the cap nut and olive on the other 'run' end of the tee.

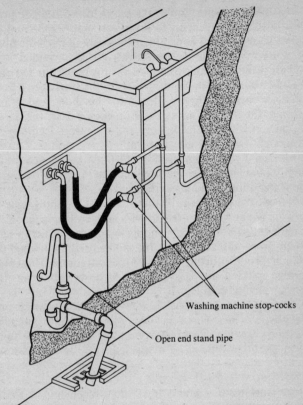

Washing machine stop-cocks

Open end stand pipe

Fig. 80 Plumbing in a washing machine

'Spring' the two cut ends of pipe into the body of the compression fitting. Smear jointing compound on to the pipe and on to the olive. Remove the clothes pegs. Push the olives and cap nuts up to the body of the tee, and screw up to complete the joints as describe in the previous chapter.

Cut a piece of 15 mm copper tubing to the length required to reach the position of the washing machine stop-cock and fit this into the outlet of the compression tee.

Washing machine stop-cocks are made with a compression inlet for connection to a copper supply pipe and an outlet designed for connection to a washing machine hose. They normally have an attractively styled shrouded head and are provided with a base for screwing to the kitchen wall.

Remove the cap nut and olive from the stop-cock's inlet. Push the branch supply pipe into this inlet and mark, through the screw-holes in the stop-cock's base, the points at which it will need to be screwed to the wall. Remove the stop-cock. Drill and plug the wall at these points.

Finally, fit the compression inlet to the branch water supply pipe and screw the stop-cock to the wall. Reinstate the water supply and check for leaks.

Repeat the process with the other water supply pipe to provide hot and cold water supplies to the washing machine. If the kitchen hot and cold supply pipes are parallel to each other, some difficulty may be experienced in taking the branch from the further pipe past the nearer one. This could involve making two easy bends in the branch from the further pipe to enable it to pass.

An easy way out of this difficulty is to use a compression *cross-over coupler*. This is, in effect, a longer-than-usual compression coupling which incorporates a bend in its body to permit another pipe to be by-passed.

The Kontite Thru-flow valve

If the washing machine is near the kitchen hot and cold water supply pipes an alternative way of providing water is with *Kontite Thru-flow valves*. These are combined tee junctions and stop-cocks which allow the washing machine hoses to be connected directly to the main hot and cold supply pipes.

The 'tee' ends of this valve consist of *Kontite Type A* compression joints, one of which does not have a tube stop in the joint body. This makes easier the task of springing the tee in to the cut run of pipe.

To fit a Thru-flow valve drain the water supply pipe as

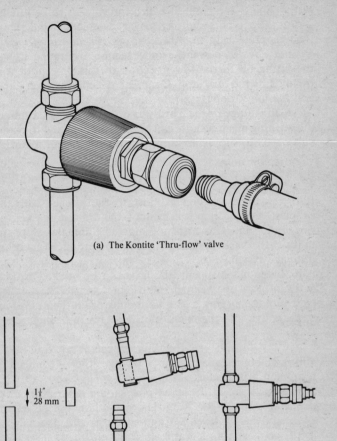

(a) The Kontite 'Thru-flow' valve

(b) Fitting the 'Thru-flow' valve

Fig. 81   The *Kontite* 'Thru-flow' valve

previously described. Cut a segment 28 mm (1 in) long out of the pipe at the point at which the valve is to be fitted.

Slip the cap nuts and olives over the cut ends and secure them as previously described. Next push *the end of the tee without a tube stop* over the upper section of cut pipe. Lower the fitting on to the lower section of this pipe until it is retained by the tube stop. Complete the compression joints and the job is done.

An advantage of the Thru-way valve is that, if required, the washing machine hose can be disconnected and the same valve used to supply an automatic dishwasher or a garden hose.

Washing machine drainage

If the washing machine is close to the kitchen sink the simplest way to drain it is to hook the drain hose over the rim of the sink. This costs nothing and eliminates all possibility of the back-siphonage that *can* be produced by other drainage systems.

If this is inconvenient, the hose can be hooked permanently into an open-ended stand-pipe fixed to the kitchen wall and discharging over a yard gully or, in the case of an upstairs flat, into the main drainage stack of a single stack system.

Always consult the Building Inspector at your local authority before making this – or any other – connection to a main waste stack.

The open end of the stand-pipe should have an internal diameter of at least 35 mm ($1\frac{1}{3}$ in) and should be 60 cm 24 in) above floor level. A trap must be provided at its base if it is connected to a waste stack but if it discharges over an external gully there is no real need for a trap.

Manufacturers of uPVC drainage systems include washing machine drainage stand-pipes in their range of equipment.

Fitting an outside tap

For gardeners and car owners an outside tap is desirable. Trailing a hose through the back door to connect to the cold tap over the kitchen sink is thoroughly inconvenient and can damage the tap.

Taking a branch water supply pipe from the rising main, through the wall, to an outside tap is a simple and straightforward job. First though, you must obtain the permission of the local Water Authority. This is likely to be granted readily enough though they will almost certainly make an extra charge on your water rate.

Decide the position against the external wall at which you want the outside tap to be fitted. Bear in mind that it must be high enough for a bucket or watering can to be placed comfortably underneath it and that the nearer it is to the rising main the cheaper and simpler the task of installation will be.

Measure the height of the proposed new tap above ground level. Measure this distance, plus the difference between the level of the kitchen floor and that of the ground outside, up the rising main. Measure a further 6 in up the rising main and mark the pipe at this point. It is from here that the branch to the outside tap must be taken.

Turn off the main stop-cock and drain from the cold tap over the kitchen sink and – if there is one – from the drain-cock immediately above the stop-cock.

Cut an 18 mm ($\frac{3}{4}$ in) section out of the rising main at the point marked and fit an equal ended 15 mm compression tee as described in plumbing in a washing machine. Make sure that the outlet of the tee is parallel to the wall and points in the direction of the outside tap.

Fit a short – say 15 cm (6 in) or 25 cm (9 in) – length of 15 mm copper tubing into the outlet of the tee. Then fit a screw-down stop-cock with compression joint inlet and outlet to the other end of this tube. Be sure to fit the stop-cock so that the arrow engraved on its body points in the direction of water flow – away from the rising main. Angle it away from the wall so that it can easily be turned on or off without grazing the fingers.

Turn the new stop-cock off. Open up the main stop-cock to re-establish the water supply. You can now proceed with the remainder of the task at your leisure.

Cut a hole through the wall at the same level as the new branch and immediately above the position at which the outside tap is to be fitted. This can be done with a cold chisel and hammer but an electric drill with a long masonry bit is preferable.

Measure the distance from the centre of the hole that you have cut through the wall to the outlet of the new stop-cock. Cut a piece of 15 mm copper tubing to this length. Cut another piece of tube long enough to pass through the hole in the wall and to project 25 mm (1 in) beyond it.

Join these two lengths of tube with a 15 mm elbow bend with compression joint inlet and outlet. Push one length through the hole in the wall and connect the other end of the other length of tube to the outlet of the new stop-cock.

Fit another compression elbow bend on to the projecting end of the pipe thrust through the wall. Make sure that its outlet points downwards.

Into this outlet fit a short length of copper tubing to reach the point at which the tap is to be fitted.

The tap itself must be fitted into a wall plate elbow with compression inlet and female iron outlet. Remove the cap nut and olive from the compression inlet and push it on to the short length of copper tube as far as the tube stop. Mark through the screw-holes in the wall plate to the wall behind.

Remove the fitting. Drill and fit plug in the wall at the points marked. Fit the compression joint inlet of the wall plate elbow on to the copper tube and screw the plate firmly to the wall to secure it.

The outside tap should be a bib-tap with a hose connector outlet and should be angled away from the wall to avoid

(a) wall plate elbow

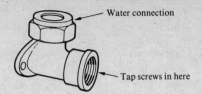

Water connection

Tap screws in here

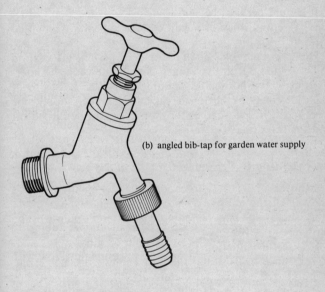

(b) angled bib-tap for garden water supply

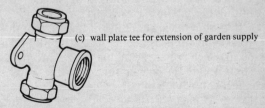

(c) wall plate tee for extension of garden supply

Fig. 82 Fitting an outside tap

grazing the fingers. Bind a turn of PTFE thread sealing tape round the male thread of its inlet and screw into the outlet of the wall-plate elbow. If – as is quite likely – it can't be screwed tightly to an upright position, unscrew it again and add one or more metal washers to its tail until an upright position is achieved.

All that remains to be done is to make good the wall and to open up the new stop-cock that you have fitted into the branch supply pipe.

It should perhaps be said that this stop-cock isn't *absolutely* essential when fitting an outside tap. Many are fitted without one.

However the provision of a stop-cock permits installation to be carried out in two stages, thus minimising the interruption of the household plumbing services. It is also of value in protecting the outside tap and its branch supply pipe from frost. In icy weather the stop-cock can be turned off and the tap drained.

An extension to the outside water supply

For those with a large garden, one outside tap fitted against the wall of the house may not be sufficient. One or more other stand-taps may be required at different points in the garden.

It is in this situation that polythene tubing, because of its resistance to frost and the long lengths in which it can be purchased, can be particularly useful.

Proceed as described for fitting an outside tap but, instead of fitting the tap into a back-plate elbow, use a back-plate tee. It is from the lower outlet of this tee that the additional garden supply pipe is taken. If you cannot find a tee of this kind to which polythene tubing can be directly connected fit a short length of copper tubing to the lower outlet and connect your length of polythene tubing to this by means of a 15 mm copper to $\frac{1}{2}$ in polythene compression coupling.

Although the polythene tubing is unlikely to be affected

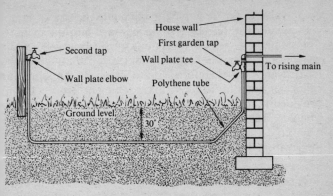

Fig. 83   An extended garden water supply

by frost it should still be laid at a depth of about 80 cm (2 ft 6 in) to avoid the risk of its being disturbed by gardening.

It can be taken underground to any point required and connected to a tap fixed to a post or to the wall of an outbuilding by means of the usual back-plate elbow.

Plumbing in a mains water softener

The way in which a mains water softener can be used to reduce a hard domestic water supply to zero hardness is described in Chapter 6. Plumbing such an appliance into the rising main may seem a somewhat daunting task yet the plumbing skills required are no different from those that are needed to fit an outside tap or to plumb in an automatic washing machine.

First read the manufacturer's instructions carefully, then ask at your local Water Authority offices about any regulations relating to the installation of these appliances.

It is possible that the manufacturers will advise the provision of a pressure reducing valve on the flow pipe supplying the softener with water. The Water Authority are likely to insist upon the provision of a non-return valve or *air break* valve in this flow pipe to eliminate the risk of brine-contaminated water being siphoned back into the main.

The main stop-cock must, of course, be closed while installation takes place. This is not a job that should be hurried and it is therefore wise to draw off three or four gallons of water for drinking and cooking before you begin. Provided that you have an indirect cold water system (see Chapter 2) there will be a reserve of 50 gal of water in the main cold water storage cistern for washing and flushing the lavatory suite.

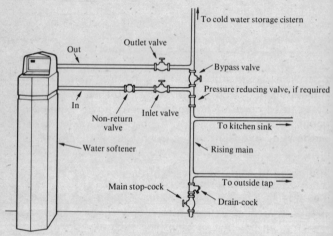

Fig. 84   Plumbing in a water softener

The flow pipe to the water softener should be taken from a point above any branch supplying an outside tap. It is also, in my opinion, better to take it from a point above the branch taken off to supply the cold tap over the kitchen sink. The reason for this is explained in Chapter 6.

Cut the rising main at the point decided upon. If a pressure reducing valve is recommended by the manufacturers it is at this point that it should be inserted into the main.

Immediately above this valve must be fitted the 15 mm

compression tee that is to take the flow pipe to the water softener. Connect a short length of 15 mm tube into the outlet of this tee and, to the other end of this tube, fit a screw-down stop-cock with compression joint inlet and outlet. Make sure that the arrow on the stop-cock body points towards the water softener.

Fit a further short length of copper tube into the outlet of the stop-cock and, on to the further end of this, connect the non-return valve that will probably be required by the Water Authority.

All that now has to be done, so far as the flow connection is concerned, is to take a length of copper tube from the outlet of the non-return valve to the compression joint inlet of the water softener. It is probable that one or more bends will need to be made in this length of pipe. These bends can be made in any one of the ways suggested in the previous chapter; with compression or capillary elbow bends, by spring bending or – since the pipe will be out of sight behind the water softener – with U-can copperbend.

Before proceeding with the return pipe from the softener, go back to the rising main. Immediately above the tee junction into which the flow pipe is connected fit another screw-down stop-cock. When the water softener is in operation this *by-pass* valve will be kept closed. However, if it should be necessary to remove the softener for servicing or repair, this valve can be opened and the stop-cocks on the flow and return pipes will be closed.

Above the by-pass valve fit into the rising main another 15 mm compression tee to take the return pipe from the softener. Fit a short length of copper tube into the outlet of this tee and, on to the other end of this, fit the return pipe's screw-down stop-cock. This time the arrow on the valve's body must point away from the water softener and towards the rising main. Take the return pipe from the water softener to connect to the inlet of this stop-cock.

Now, opening the stop-cocks on the flow and return pipes and closing the by-pass valve in the rising main will ensure that all water flowing through the rising main passes through the water softener.

The drain-hose of the water softener – which is brought into use during the flushing and recycling process – resembles the drain hose of an automatic washing machine. It should discharge into the drain, in exactly the same way as a washing machine, through an open-ended stand-pipe.

Most water softeners are provided with an overflow outlet. A pipe from this should be taken through the wall to discharge into the open air in the same way as an overflow pipe from a lavatory or main storage cistern.

All that remains to be done is to connect a 220/240 v electric supply to the time clock that prompts the regeneration process. This can be connected to an ordinary 13 amp socket outlet but water softener manufacturers usually recommend that the appliance should be wired to a fused connection unit to prevent it being accidentally switched off. A 3 amp fuse should be used.

All manufacturers supply easy to follow instructions on setting the water softener's programmer. The actual setting will depend upon the hardness of the local water supply and upon the anticipated water consumption of the family.

One final point; installing a mains water softener – particularly if the manufacturer has advised the fitting of a pressure reducing valve – will reduce the flow of water into the main cold water storage cistern. You may find it necessary to substitute a low pressure ball valve or – preferably – an equilibrium ball valve for the high pressure valve that is normally used to serve this cistern.

13

Plumbing Work in the Bathroom

The bathroom fittings of the average suburban home have, for many years, consisted of a bath, a wash hand basin and a lavatory suite, perhaps with the last of these in a separate – usually adjoining – room.

Changes are taking place though. Many modern bathrooms now have a shower fitted over the bath. And, limited space – particularly in the conversion of large, old houses into self-contained flats – has resulted, in many instances, in a shower cabinet replacing the bath altogether. Such a cabinet need occupy a floor area no greater than a yard square.

Bidets too, regarded before the Second World War – by those who had heard of their existence – as a not-quite-respectable Continental extravagance, are increasingly to be found in British bathrooms. I think it possible that an independent shower and a bidet will be standard equipment in the well-appointed bathroom of the future and that the *very* well appointed bathroom may well have its own sauna compartment – perhaps replacing the traditional sit-down bath.

Renewing a bath

Be that as it may, there is no doubt that the bath will be with us for many years to come, and so replacing an old bath is a task with which any householder may be faced.

Baths may be made of enamelled cast iron, of enamelled pressed steel or of a variety of plastic materials including glass-reinforced plastic and acrylic sheet. They have widely differing characteristics which should be carefully considered when choosing a replacement.

The existing bath will almost certainly be made of enamelled cast iron – the traditional bath material. This is strong and hardwearing (although it can chip) but baths made of this material are very expensive and very heavy. They are consequently quite incapable of 'one man' installation. Enamelled cast iron is a good conductor of heat, which is a positive disadvantage as the warmth of the water is conducted away.

Pressed steel baths are appreciably cheaper and lighter in weight, but the thin material makes them very liable to accidental damage in storage and installation. However one manufacturer claims to have overcome this objection by producing a 'supersteel' bath. This is made of enamelled sheet steel 50% thicker than that normally used. Baths of this material are however only half the weight of a cast iron bath of the same size and they are, of course, markedly cheaper.

Baths made of acrylic or other plastic material have achieved growing popularity in recent years. They are tough and hard wearing and are sufficiently light in weight to be capable of being carried upstairs and installed by one man working alone. They are available in a variety of colours and the colour extends right through the material of which they are made. Surface scratches can be polished out.

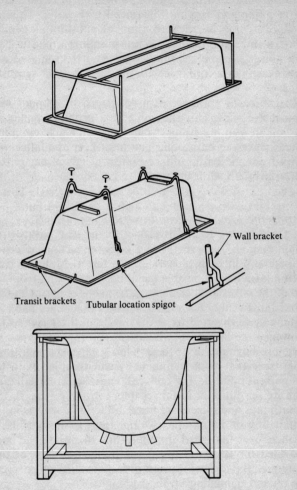

Fig. 85   Frames for acrylic plastic baths

Early plastic baths acquired something of a bad name for creaking and sagging when filled with water. This has been overcome by the provision of substantial padded wooden or metal 'cradles'. When installing a plastic bath you must assemble the cradle and fix it to the wall and floor exactly in accordance with the manufacturer's instructions.

Plastic baths are more easily damaged by extreme heat than metal ones. This shouldn't be a problem in normal use but, if you are using a blow lamp for any plumbing operation in the bathroom, you must keep the flame well away from the bath; and never rest a lighted cigarette – however briefly – on the rim!

Fitting the taps, waste and overflow

Fit the taps, waste and overflow to the new bath before attempting to remove the old one. You can use either individual $\frac{3}{4}$ in bath taps or a bath mixer. Many modern bath mixers have a flexible metal hose projecting from the top of the mixer and terminating in a shower sprinkler that is fixed to a wall bracket over the bath. Mixed hot and cold water can be diverted from the nozzle of the mixer to the shower at the flick of a switch.

Before buying one of these – fairly expensive – fittings, make sure that your plumbing system complies with the design requirements for conventional shower installation that are set out later in this chapter.

Bath taps and mixers are fitted in the same way as in the installation of a sink unit (see pp. 190–1). Slip flat plastic washers over the tails before inserting them into the holes provided for them. Then – since baths are made of thin material – slip top-hat washers over these tails before screwing on and tightening up the back nuts.

Bed the flange of the bath waste on to non-setting mastic smeared liberally round the outlet hole provided for it. Slip

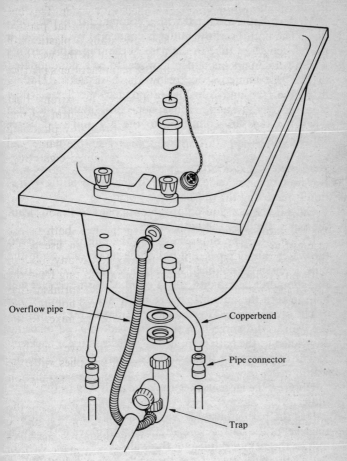

Overflow pipe

Copperbend

Pipe connector

Trap

Fig. 86   The connections to a modern bath

a flat plastic washer over the tail of the waste before screwing up the back nut.

Modern baths are fitted with a combined trap and overflow. The bath overflow outlet is connected by means of a flexible length of tube to the base of the trap. This permits a trap with an adequate depth of seal to be accommodated in the very limited space below the bath outlet. Having fitted the trap and overflow, disconnect the flexible pipe from the overflow outlet. Final connection should take place when the taps are connected to the hot and cold water supply pipes.

Removing the old bath

It is now time to tackle the most difficult part of the operation – removing the old bath.

Cut off the hot and cold water supplies to it and drain the pipes. Remove the bath panel.

Unscrew the cap nuts connecting the tails of the taps to the water supply pipes and pull these pipes away from the tails. Even in the confined space behind the foot of the bath this shouldn't be too difficult. A cranked basin spanner will help to turn the nuts if they are really inaccessible.

The overflow pipe from an old bath may be taken through the wall to discharge into the air outside instead of being connected to the bath trap. If this is the case use a hacksaw to cut off this pipe flush with the internal surface of the wall.

Unscrew the large nut securing the bath trap to the waste outlet.

Has the bath adjustable legs? If so adjust them to lower the level of the bath before attempting to move it. This will reduce the risk of damaging the wall tiles.

Pull the bath away from the wall. You will find that it is extremely heavy. For this reason it is better not to attempt to move it from the bathroom in one piece – even if you have strong helpers.

As cast iron is brittle, the bath can be broken into manage-

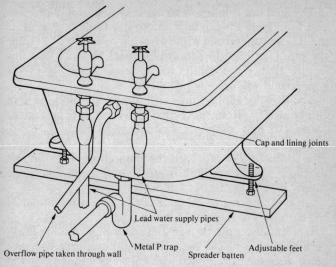

Cap and lining joints

Overflow pipe taken through wall

Lead water supply pipes

Metal P trap

Spreader batten

Adjustable feet

Fig. 87   At the foot of an old bath

able pieces with a club hammer. Wear goggles to protect your eyes as you do this and drape the bath with a blanket to stop pieces flying about the room.

Fitting the new bath

Having removed the old bath, move the new one into position. Note the position of the water supply pipes in relation to the tails of the new bath taps. If you are *very* lucky they may come to exactly the right position for connection. It is much more likely though that you will have to remove the swivel tap connectors and either lengthen or shorten the supply pipes to meet the new positions of the taps.

This can be a very difficult job in the dark and confined space at the foot of a bath.

The simplest solution is to cut back the supply pipes to end 14 in from the tap tails and to use 14 in lengths of 22 mm U-can copperbend with swivel tap connectors to bridge the gap. Don't cut the supply pipes back by more than this length. If they are a little too long the flexibility of the copperbend will accommodate this.

The existing supply pipes will almost certainly be $\frac{3}{4}$ in Imperial size. Remember that adaptors are needed to connect 22 mm compression couplings to them.

Make your connections in logical order; first the supply pipe to the further tap, then the connection of the flexible overflow pipe to the bath overflow outlet and finally the supply pipe to the nearer tap.

All that remains to be done is to connect the outlet of the bath trap to the branch waste pipe.

Incidentally, a plastic trap and waste pipe should always be used with a plastic bath. Thermal movement will take place when the bath is filled with hot water and the material of which it is made could be damaged by a rigid metal trap and waste pipe.

Fitting a shower

Showers have many advantages over baths. They save time, water and fuel bills – manufacturers claim that four or five satisfying showers can be obtained from the same volume of water that is used in just one bath. They are more hygienic – the bather is not sitting in his own dirty water! They are safer for the very young and for the elderly and there is less cleaning up to be done after use. An independent shower fitted into its own cabinet takes up far less floor space than a conventional bath.

There are certain, quite definite, design requirements that must be met if a shower is to be supplied with hot water from a cylinder storage hot water system.

The first, and perhaps the most important, is that the

hot and cold water supplies to the shower must be under equal pressure. The hot supply from a cylinder storage system derives its pressure from the main cold water storage cistern. The cold supply to the shower must therefore be taken from the same cistern – *not* direct from the main.

It is illegal to mix, in any plumbing appliance, water originating from a storage cistern with water from the main. It is also quite impracticable because of the great difference

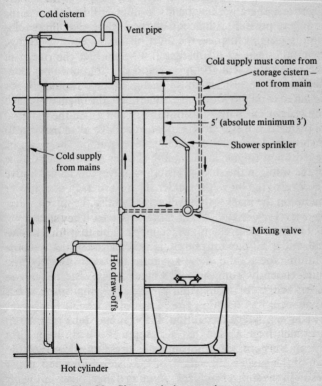

Fig. 88   Shower design requirements

in pressure between a main and a cistern supply. If attempted it would be found that the shower will supply only cold water until the cold supply is almost turned off, and would then, quite suddenly, run scaldingly hot.

The second essential is that water pressure at the shower must be adequate. This will be assured if the level of the base of the cold water storage cistern is at least 5 ft (1.5 metres) *above* the level of the shower sprinkler. The absolute minimum 'head' that will give an acceptable shower is 3 ft (about 90 cm) but this will be effective only if pipe runs are short and there are few bends to impede flow.

Where pressure is inadequate the cheapest and best solution is usually to raise the level of the cold water storage cistern. Construct a substantial wooden platform for it in the roof space, cut the supply and distribution pipes connected to the cistern and – using either compression or soldered capillary joints – insert new lengths of pipe to reach the cistern in its new position.

If this is not practical a more expensive alternative is to install an electrically operated shower pump. These can be relied upon to give an adequate shower even if the surface of the water in the storage cistern is only a few inches above the level of the shower sprinkler. There is, in fact, one shower pump on the market that will actually raise the mixed water to a level *above* that of the storage cistern if required.

The final requirement, which may not be absolutely essential under all circumstances, is that the cold water distribution pipe to the shower should be taken directly and independently from the cold water storage cistern – not as a branch from a distribution pipe serving some other appliance.

This is a safety precaution. If the supply pipe is taken as a branch from a pipe that also supplies a wash basin or a flushing cistern, the drop in water pressure when the cold tap of the basin is turned on, or the cistern is flushed, could result in the shower suddenly running dangerously hot while in use.

All showers of the kind so far described must have a mixing valve of some kind in which the hot and cold supplies of water can be mixed to an acceptable temperature. The simplest kind of mixer consists of two stop-cocks or taps connected together. This arrangement is provided by the bath/shower mixers already described. The user adjusts the two tap or stop-cock handles until water at the required temperature is obtained.

The manual shower valve – fitted as standard in most independent shower cabinets – offers an improved means of mixing. The two streams of water are mixed in a single valve. Temperature, and sometimes volume, are regulated by turning a large knurled knob.

A further improvement is offered by the *thermostatic mixing valve*. Like the manual valve this is operated by a single control knob but a thermostatically operated device inside the valve ensures a flow of water at a preset temperature despite minor pressure fluctuations in either the hot or the cold water supply. The principal value of these appliances is to hotels, schools and sports complexes where a number of showers may be supplied with hot and cold water from the same source. In the home they can provide an additional safety factor and make it unnecessary to run an independent cold distribution pipe from the cold water storage cistern to the shower.

Thermostatic mixing valves have their limitations though. They cannot – even if this were legally possible – accommodate the great difference in pressure between a mains cold water supply and a cistern hot water supply. Nor can they *increase* pressure on either side of the shower. When pressure falls on one side, that on the other side is reduced to equal it. If, for instance, a shower is already operating on minimal hydraulic head, any reduction in pressure in the water supply on either side of the shower will simply result in the shower drying up until pressure is restored.

Showers supplied from instantaneous heaters

Most manufacturers of multipoint gas instantaneous heaters provide for the connection of a shower. Usually though, these only work if no hot water is being drawn off from any other point in the system.

Electric instantaneous shower heaters have become increasingly popular in recent years. They only need connecting to a 15 mm branch supply pipe teed from the rising main and to a suitable electric supply. Temperature control is obtained in most cases simply by increasing or decreasing the flow.

This kind of shower can be quickly and easily installed where there is no cylinder storage hot water system or where the cold water storage cistern is situated at too low a level to give adequate pressure.

The rate of flow, although adequate, is appreciably less than that from a conventional shower and, in hard water areas, trouble can arise from scale formation within the appliance.

The sauna

Saunas are not, strictly speaking, plumbing installations but – used in conjunction with a shower – they can provide a relaxing and refreshing alternative to the 'good soak' that many people enjoy in a bath.

Indoor saunas consist of an insulated pine-wood cabin which can occupy a floor area as small as 1 × 1.5 metres (3 ft 9 in × 4 ft 6 in). They are heated by a special sauna stove in which rocks are heated by banks of electric elements positioned beneath them. The stove is thermostatically controlled to give any predetermined air temperature within the cabin.

Gas and wood-fired sauna stoves are also available.

Because the heat is dry, very high temperatures can be tolerated – 85°F to 90°F is quite usual. To use a sauna take a warm shower before entering the cabin. After about fiteen minutes in the dry heat take a cold shower to close the pores and then re-enter the cabin for a further fifteen to twenty minutes.

Towards the end of that period water can be poured – fairly cautiously – on to the hot stones of the stove. Instantly the dry heat is converted to a damp heat to produce the 'heat shock' beloved by sauna addicts.

This process can be repeated as often as desired. After a final cold shower it is wise to relax on a bed or couch for half an hour or so.

Saunas should not be taken – without medical advice – by people with respiratory or circulatory problems but, for anyone in normal health, a regular sauna can be a refreshing and invigorating experience – just the thing after two or three hours hard work in the garden.

At least one sauna manufacturer supplies detailed instructions for d-i-y installation.

The wash basin

Saunas are – and are likely to remain – an exotic exception in a British bathroom. A wash basin, on the other hand, is an essential fitting in every bathroom, and is now often fitted in a bedroom and, where the lavatory is separate from the bathroom, in that compartment too.

Wash basins may be made of ceramic material, of enamelled pressed steel or of plastic. Basins made of the last two materials are most likely to be found fitted into 'vanity units'.

Ceramic basins are most possible and are available as *wall hung* or *pedestal* models. Wall hung basins take up no floor space and can be fitted at the level most suitable for those most often using them. They are usually fitted with the rim about 80 cm (32 in) above floor level but – in a nursery for instance – a lower level may be preferred.

Pedestal basins are considered more attractive and conceal the water supply and waste pipes. They also have the advantage of providing extra support where the basin is to be positioned against a weak partition wall.

Modern pedestal basins are never supported by the pedestal alone. Concealed brackets or hangers fixed to the wall provide extra support, which is particularly useful when the basin is being fitted or when the pedestal has to be removed to gain access to the water supply pipes or the waste outlet.

Basins may be provided with individual $\frac{1}{2}$ in hot and cold taps or with a basin mixer. A basin mixer should be used only where there is an indirect cold water supply (see Chapter 2).

Some modern basin mixers incorporate a 'pop-up waste'. Pressure on a knob positioned between the two handles of the mixer raises the waste plug to empty the basin. This eliminates the rather unsightly plug and chain (and also makes it impossible for the plug to be stolen or 'borrowed' – of greater interest to hotel and restaurant proprieters than to domestic users!).

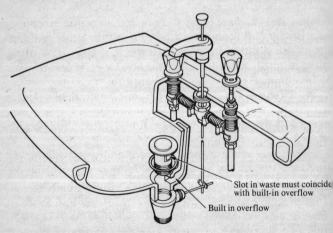

Slot in waste must coincide with built-in overflow

Built in overflow

Fig. 89   *Deltaflow* basin mixer with pop-up waste

Basin taps and mixers are fitted in a similar way to those serving sinks and baths. As a ceramic basin is relatively thick, a 'top-hat' washer is not usually required. Slip flat plastic washers over the tails before screwing on the back-nuts. These should be tightened sufficiently to hold the taps securely in the position required but should not be over-tightened. Be careful, as ceramic basins are easily damaged.

Water supply to the taps should be taken in 15 mm copper tubing connected by means of swivel tap connectors or cap-and-lining joints. Supplies to pedestal basins are usually taken up vertically inside the pedestal. This involves making two easy bends in the supply pipes to enable them to branch out to connect to the tap tails. This is, once again, a con-cealed position in which 15 mm U-can copperbend can be very useful.

Ceramic basins usually have a built-in overflow. A *slotted waste* must therefore be bedded down, on mastic, into the hole provided for it in the basin. Before slipping a washer on to the tail of the waste and screwing up the back-nut, make sure that the slot in the waste coincides with the outlet of the built-in overflow.

Wash basins are usually fitted with chromium-plated or plastic bottle traps. For a wall-hung basin these are more attractive than the conventional U trap, and take up less of the limited space inside the pedestal.

The ultimate destination of the waste pipe will depend upon whether the house has two-pipe or single-stack drain-age. Your local authority's building inspector should always be consulted before making this – or any other – connection to a main drainage stack.

The basins of vanity units are plumbed in in exactly the same way, except that, you will need 'top hat' washers on the tap tails as they are made of a relatively thin material. It is probable too that the basin of a vanity unit will not have a built-in overflow. The overflow will be taken, like that of a sink units, by means of a flexible pipe to connect to the waste outlet above the trap.

Changing the taps of baths and basins

Anyone who has successfully fitted a new bath or basin might imagine that removing the old taps from an existing bath or basin would be a simple and easy job.

This is not always the case. It is usually easy enough to disconnect the tap tails from the water supply pipes but unscrewing the back nuts that secure the taps to the bath or basin can be quite a different matter.

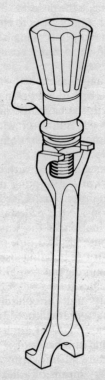

Fig. 90 Using a cranked bath or basin spanner

The back-nuts are inaccessible and are often firmly fixed by long-dried-out putty, by hard water scale and by corrosion.

The application of penetrating oil followed by the use of a cranked bath or basin spanner may release them. However if you are planning to replace basin taps it is usually quicker in the long run to disconnect the tap tails from the water supply pipes and the trap from the waste outlet. Then lift the basin off its brackets or pedestal and turn it upside down on the floor. In this postion you will find it much easier to turn the nuts without damaging the basin.

As you can't turn a bath upside down you will have to disconnect its waste outlet and the water supply pipes and pull it forward to give yourself more room to work in.

Having removed the old taps and fitted the new ones you may be faced with another unpleasant surprise when you try to connect their tails to the water supply pipes.

The tails of modern taps are marginally shorter than those of older ones, so it is possible that the swivel tap connectors will no longer meet and screw up on to the tails. Usually water pipes have enough 'give' in them to permit this difference to be accommodated. However, manufacturers of compression fittings make an extension piece specially designed to bridge the gap where necessary.

The bidet

The bidet is a specially designed, low level wash basin used for washing the lower parts of the body. (It can also serve a useful secondary purpose as a foot bath!) Bidets are gaining in popularity and are certainly well displayed in showrooms and in illustrated bathroom equipment catalogues.

There are two kinds of bidet, one of which is much more expensive – and can be much more difficult to fit – than the other. It is very important that the purchaser and installer should be able to recognise the difference.

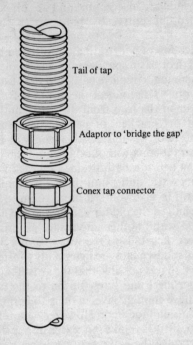

Tail of tap

Adaptor to 'bridge the gap'

Conex tap connector

Fig. 91 'Bridging the gap' between water supply pipe and tap tail

The simpler and cheaper kind is the *over-rim supply bidet*. Water supply is from two pillar taps or – more likely – a mixer, which delivers water into the bidet from above the level of the rim, in exactly the same way as the taps of a wash basin.

Plumbing-in is done in the same way as a wash basin. The hot and cold water supply pipes are taken as 15 mm branches from the 22 mm or $\frac{3}{4}$ in hot and cold supply pipes to the bath. If these are of the old, Imperial $\frac{3}{4}$ in size, remember that you can't fit a 22 mm to 15 mm reducing tee into them without the appropriate adaptors.

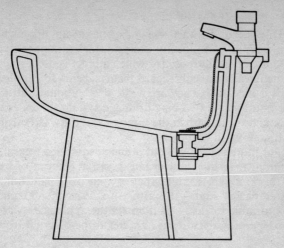

(a) Over-rim supply bidet – resembles a wash basin in all respects

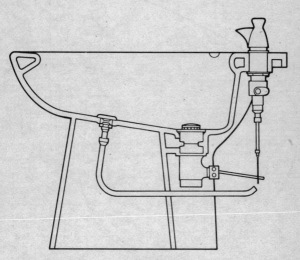

(b) Rim supply bidet with ascending spray

Fig. 92 'Over-rim' and 'rim supply' bidets

Disposal of the waste depends, once again, upon whether you have a two-pipe or a single-stack drainage system. Because of the purpose for which a bidet is used, plumbers sometimes mistakenly regard them as being soil appliances. This is incorrect. They are waste or ablutionary fittings. If the house has a two-pipe drainage system the bidet waste should be dealt with in the same way as the wastes from the bath and wash basin.

The more expensive kind of bidet is referred to as having a *through rim supply with rising spray*. The rim of a bidet of this kind is not unlike that of a lavatory pan. Inflowing water passes round this rim – to warm it and make it comfortable for use – as it enters the appliance. A control knob permits the stream of water to be diverted from the

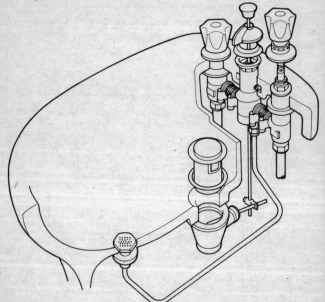

Fig. 93   *Deltaflow* bidet set with ascending spray and pop-up
waste

rim to a submerged spray which directs the water to those parts of the body to be cleansed.

It is the rising spray that complicates the installation. Submerged water inlets are always regarded with suspicion by Water Authorities because of the risk of contaminated water being siphoned back into the main. The use to which a bidet is put makes it doubly important that there should be no possible risk of this occurring.

Always consult your own local Water Authority before purchasing or installing this sort of bidet. Regulations may

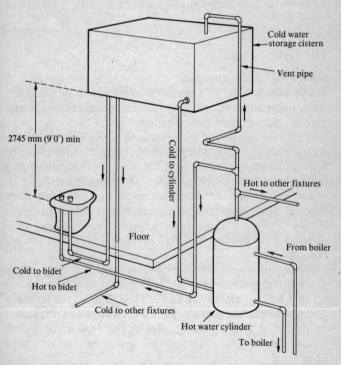

Fig. 94   Design requirements for rim supply bidet with ascending spray

vary slightly from one part of the country to another.

Usually though, they will require the cold water supply to the bidet to be taken in a separate 15 mm distribution pipe directly from the cold water storage cistern – not as a branch from any other cold water distribution pipe. Similarly the hot water supply to the bidet will have to be taken in a separate 15 mm distribution pipe from the vent pipe above the hot water storage cylinder.

A final requirement is likely to be that there must be a vertical distance of at least 2.75 metres (9 ft) between the base of the cold water storage cistern and the level of the bidet inlet.

The lavatory suite

Renewing a lavatory pan

Many homes have two lavatory suites – one of them up-stairs, in or adjacent to the bathroom, the other downstairs, perhaps approached from outside.

Renewing the pan of an upstairs lavatory suite is usually a simple enough task. The pan will be secured to the boarded floor of its compartment with brass screws. The pan outlet connects to the branch soil-pipe with a flexible mastic joint.

To remove the old pan, disconnect the flush pipe from the pan's *flushing horn*. Unscrew and remove the brass screws. Pulling the pan forward while, at the same time, moving it to and fro, will free the outlet from the branch soil pipe and permit it to be removed.

Removing a ground floor lavatory pan can be a much more difficult task. The base of the pan will probably be set in a bed of cement, on to the solid floor of the lavatory compartment. The pan will have an S trap outlet connected by means of a rigid cement joint to the protruding socket of a stoneware branch drain. This socket will be just above floor level.

To remove the pan you will have to break it with a club hammer. Disconnect the flush pipe. Then use your hammer

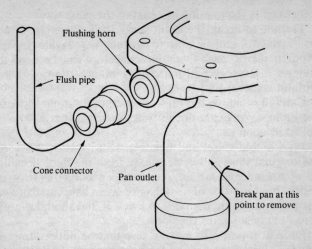

Fig. 95   Removing an old lavatory pan with cemented-in S trap
outlet

to break the pan outlet, situated behind the trap and just above the connection to the drain socket.

Drive a cold chisel under the base of the pan to free it from its cement bed, and pull it forward out of the way.

You will be left with the sand and cement base on to which the pan was set and the pan outlet protruding jaggedly from the drain socket.

First, clear the socket. Stuff a bundle of rags or newspaper into the branch drain to prevent pieces of the lavatory outlet or the cement jointing from falling into the drain to block it.

Attack the pan outlet with a small cold chisel and hammer. Work carefully and patiently, keeping the blade of the chisel pointing towards the centre of the drain socket. If you can break the pan outlet right down to its base at one point you will usually find that the remainder of the outlet can be prised out fairly easily. When you have removed the pan outlet tackle the jointing material in the same way.

Try not to break the drain socket. (Don't panic if you do – you will still be able to fit your new lavatory pan. Modern plastic push-fit drain connectors can be used to connect a lavatory pan outlet to a branch drain with or without a socket.)

Carefully remove the rags or newspaper 'stopper' from the drain, taking care not to spill the debris that will have collected upon it.

With a cold chisel and hammer, remove every trace of the original sand and cement base from the floor, leaving a smooth and level surface. As it has been found that stresses caused by the setting of a sand and cement base can damage the pan you should not need to renew it, but should simply screw the new pan to the floor.

Place it in position, making sure that its outlet projects centrally into the socket of the branch drain. The distance between the outside surface of the pan outlet and the internal wall of the socket should be equal at all points.

When the position of the pan has been established mark through the four screw-holes in its base to the floor beneath with a ball-point refill. Then draw a pencil line on the floor, round the base of the pan, to enable you to replace it in exactly the same position when required.

Remove the pan. Drill the floor at the four points that you have marked and plug the holes to take fixing screws.

The new pan can be connected to the branch drain either with a mastic filler and building tape joint as described in Chapter 5 ('Leakage from lavatory drain connection') or with a patent plastic push-on joint referred to in the same chapter.

If you choose the push-on joint this must be fitted before the pan is finally screwed into position. The mastic filler joint is made after this has been done.

Place the pan carefully into the pencil-marked position and probe through the screw holes to check that your floor plugs are in exactly the right position. Slide lead washers over

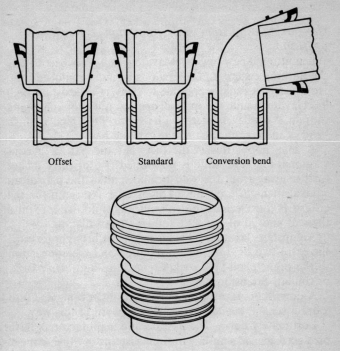

Offset   Standard   Conversion bend

Fig. 96   *Multikwik* drain connectors

the brass screws before inserting them into the holes and screwing them down. These washers will prevent the ceramic surface of the pan from damage as you drive the screws home.

Check, with a spirit level, that the pan is set dead level. If it is not, loosen the screws on the lower side and pack strips of lino or slivers of wood under the base as necessary.

Converting a lavatory suite from high to low level operation

When fitting a new low-level lavatory suite it is best to buy the flushing cistern and the pan as a single, matched unit. However it is usually possible to convert an existing high level lavatory suite to low level operation without changing the pan. A low level cistern, with its short flush pipe or 'flush bend', can be connected to the existing pan.

A snag about this is the fact that the pan of a conventional low level suite is positioned two or three inches further away from the wall behind it than the pan of an equivalent high level suite. This is to make room for the cistern. If you overlook this you will find it impossible to raise the seat and lid of the suite fully after installation.

An upstairs lavatory pan can usually be brought forward the extra two or three inches required and the connection to the branch soil-pipe socket remade using a *Multikwik*, or similar, extension piece.

A downstairs lavatory pan, with its outlet cemented into a drain socket, cannot be brought forward in this way.

Fortunately during the past decade manufacturers have produced special slim-line flushing cisterns or 'flush-panels' which – in nearly every case – permit conversion of a high level suite to low level operation without the need either to move or renew the lavatory pan. The first on the market was the *Fordham Flushpanel* (Fordham Plastics Ltd, Wolverhampton). This cistern is only 11 cm ($4\frac{1}{2}$ in) from front to back and can be connected to a pan with its flushing horn inlet only 15 cm ($5\frac{3}{4}$ in) from the wall behind. A conventional low level cistern needs a 20 cm (8 in) space.

Glossary

Acorn fittings	Plastic push-fit connector made of polybutylene, immune to de-zincification and corrosion. Withstands high temperatures and pressures. An alternative to compression and capillary joints for connecting copper tubing.
Actual capacity	The capacity to the invert of the overflow or warning pipe – about 115 mm ($4\frac{1}{2}$ in) from rim.
Air lock	A bubble of air in a water distribution pipe preventing the flow of water.
Baffle board	Prevents a sudden flow of water from passing straight through a septic tank from inlet to outlet.
Ball float	A copper or plastic float, not necessarily a 'ball' attached to the arm of a valve. It moves up and down with the water level

in a tank, cutting off the water supply at a predetermined level.

Ball valve	A float activated valve used for maintaining a constant level of water in a storage or flushing cistern.
Base exchange or ion exchange	The chemical process by which a mains water softener works.
Bib tap	A tap with a horizontal inlet. Usually provides garden water supply.
Bitumastic paint	Waterproofing paint which retains its flexibility when set.
Boss white	Jointing compound.
Burr	Rough patches on tube ends. produced by cutting the tube.
Branch water supply pipe	Supplies the cold tap over the kitchen sink which provides the household's drinking water and must be supplied direct from the rising main.
Calgon	Commercial name for sodium hexametaphosphate – a water softening chemical.
Capstan head/handle	Tap handle.
Catch pit	Used in draining effluent from a septic tank.
Cesspool	A watertight underground chamber designed to contain sewage until it can be pumped away and disposed of. Constructed of brick, concrete, or glass-reinforced plastic.
Clamp seal	Rubber or plastic washer which is slipped over the waste outlet

beneath the sink above the clamp that supports the waste disposal unit.

Cistern
A water storage vessel that is 'open' to the atmosphere.

Cold water storage cistern
Intended to supply bathroom cold taps and W.C. as well as a cylinder storage system. Actual capacity should be 50 gals (227 litres).

Communication pipe
The pipe that connects the water main to the Water Authority's stop-cock.

Compression joint
A method of jointing lengths of copper tubing and connecting them to other fittings.

Continuous siphonage
The failure of the siphon to 'break' after the cistern has been flushed. The cistern does not refill and water continues to flow down the flush pipe. Pulling the chain again breaks the siphon and the cistern fills.

Cross-over coupler
A longer than usual compression coupling which incorporates a bend in its body to permit another pipe to be by-passed.

Croydon ball valve
A valve closed by a washered metal plug or piston. The plug moves vertically in the valve body and when open the water splashes noisily into the cistern via two channels built into the sides of the body. Used for cattle troughs and allotments.

Cup and cone joint	Joint for use with gas or waste pipes.
Cylinder	Closed storage vessel, usually of copper, for hot water.
Cylinder – Self-priming indirect	No separate feed and expansion tank. Has inner cylinder which also serves as heat exchanger.
Cylinder storage hot water system	A hot water supply for the whole house. Can be operated by any fuel or by a combination of fuels. May provide hot water only or form part of a hot water supply/central heating system.
Dead legs	Lengths of pipe between the storage cylinder and the hot water taps in which expensively heated water will cool after hot water has been drawn off.
Damp proof course (dpc)	A horizontal barrier built into a wall to prevent damp rising above a certain level.
Deep well (bore well)	Taps water below the first impervious stratum. Usually bored with water drawn up through a metal pipe.
Depth of seal (in a trap)	The vertical distance between the overflow outlet of the trap and the upper part of the bend at its base.
Dezincification	An electro-chemical action which affects brass fittings due to the corrosive nature of water in certain areas.
Dead soft temper tubing	Flexible tubing for underground water supply pipes and

microbore central heating installation.

Diaphragm ball valve   A valve where the float arm pushes a small metal or plastic plug against a large rubber diaphragm to close the nozzle of the valve. The valve is protected from scale and corrosion because its only moving part – the small plug is shielded from the water by the rubber diaphragm.

Direct cold water system   A system in which all cold water draw-off points (including the lavatory flushing cistern and the cold taps over the bath and wash basin) are taken direct from the rising main.

Direct hot water system   A cylinder storage hot water system in which the water in the cylinder is heated 'directly' by passing through the boiler.

Distribution pipe   A water supply pipe taken from a main cold water storage cistern.

Drain-cock   A tap, operated by a spanner, provided at the lowest points of hot water and central heating systems to permit them to be drained. Also often fitted above the householder's main stop-cock to permit the rising main to be drained.

Drain rod   Aid for unblocking drains.

'Ears'   Part of tap handle.

Easy bend   A bend in a pipeline with a

wide radius. Easy bends can be made in copper tubing with the aid of a bending spring. An easy bend should connect the soil-pipe to the underground drain.

Electrolytic corrosion
Corrosion resulting from the use of two dissimilar metals (e.g. copper and galvanised steel) in one plumbing system.

Equilibrium ball valve
Used to maintain even pressure where mains water pressure fluctuates.

Feed and expansion tank
Small open storage cistern used to supply the primary circuit of an indirect hot water system or central heating system and to provide accommodation for the expansion of the water in the system when heated.

Finger-wiped joint or taft
Soldered joint unsuitable for pipes carrying water under pressure.

Flushing horn
Inlet, at rear of lavatory pan, through which flushing water enters.

Force cup/sink waste plunger
Rubber or plastic cup designed to fit over a sink or basin waste, used in clearing a blockage.

Free outlet water heater
Water heater with free outlet and control valve on the inlet of the heater, permitting direct connection to the rising main. Operated by gas or electricity to provide hot water at sinks or basins.

Frost-stat	A thermostatically controlled device that switches the central heating on when the temperature drops to predetermined level.
Gate valve	A valve used to control or stop the flow of water in central heating circuits or distribution pipes. The waterway is closed by a metal bridge or 'gate'.
Gland nut	The nut on a tap which compresses the gland packing to stop water getting out of the tap by way of the tap spindle.
Gland packing	Material packed around a tap spindle to prevent water escaping. Knitting wool and Vaseline make a satisfactory d-i-y replacement.
GRP chambers	Drain inspection chambers made of fibre-glass reinforced plastic (GRP).
Guard pipe	A pipe (usually a 150 mm drain pipe) sunk into the soil to protect and give access to the Water Authority's stop-cock. The guard pipe has a hinged metal cover.
Gunmetal	A corrosion resistant alloy of copper and tin.
Half hard temper tubing	Rigid copper tubing for above ground domestic hot and cold water supply.
Hard water	Water containing chemicals which prevent soap from dissolving freely. It produces scale

or fur in kettles and boilers.

Hardness (of water)

Measured in terms of the equivalent of calcium carbonate in the water in parts per million (ppm).

Hardness, degrees of

To convert degrees of hardness to ppm of calcium carbonate, multiply by fourteen.

Head gear

The upper part of a tap, gate valve or stop-cock which can be unscrewed to permit servicing.

Hopper head

Device in older houses to aid discharge of first floor bath and basin wastes.

Hydraulic head

The vertical distance between a tap or other water outlet (e.g. the sprinkler of a shower) and the cold water storage cistern supplying it with water. The greater the hydraulic head the greater will be the water pressure at the outlet.

Immersion heater

An electric element, often thermostatically controlled, used to heat water in a hot water storage cylinder or a kettle.

Indirect cold water system

A cold water system in which only the cold tap over the sink is taken directly from the rising main. Bathroom and lavatory cold water supplies are taken via distribution pipes from a main cold water storage cistern.

Indirect hot water system

A hot water system, often used in conjunction with central

heating, in which the water in the cylinder is heated by a closed coil or heat exchanger situated within the cylinder and connected by flow and return pipes to the boiler.

Intercepting trap
Prevents sewer gases entering the domestic drainage system. Used in pre-war period.

Jumper
Washered valve in a tap.

Key
To roughen a surface so that another material will stick to it.

Lagging
Insulating material for pipes, cylinders and cisterns.

Magnetite
Black iron oxide sludge, a product of internal corrosion. Clogs the bottom of radiators, obstructs pipes and causes pump failure.

Mains pressure
Any tap or plumbing fitting supplied with water direct from the mains supply is at mains pressure. Other fittings are at storage tank pressure.

Marley collar-boss
Device which prevents the fouling of waste outlets while permitting them to discharge into the main stack at the same level as the lavatory outlet.

Mastic filler, non-setting
A sealing compound which retains its flexibility.

Micromet crystals
Phosphates of sodium and calcium that stabilise the chemicals which cause scale to form when water is heated.

Mixers
Valves designed for mixing a

supply of hot and cold water. Bath and basin mixers are simply two taps with a common spout. Shower mixers are manual or thermostatic. Sink mixers have separate channels within the mixer body for the hot and cold water. The two streams mix in the air after leaving the spout. Only sink mixers have the hot side supplied from a cylinder storage system and the cold side direct from the main.

Overflow	Warning pipe which discharges into the open air in a position where any overflow will attract attention.
Nominal capacity	The amount of water that a cistern would hold if filled to the brim.
P-trap	Trap with a horizontal outlet.
Packaged plumbing system	A cylinder storage hot water system (direct or indirect) in which the cold water storage cistern and the hot water cylinder are brought into close proximity to form one compact unit.
Pegged	When the jumper in a tap is fixed so that the jumper can be turned but not withdrawn.
Pillar taps	A tap with a vertical inlet. Used in modern sinks, basins and baths.
Pipe clips	Support and secure copper tubing to walls.

Plumbers solder

Two parts lead to one part tin with a melting point of 230.

Primary circuit

The closed circulation between the boiler and cylinder of an indirect hot water system, supplied from its own feed and expansion tank. Any central heating system must be connected to the primary circuit.

PTFE thread sealing tape

A plastic sealing tape bound round the thread of a screwed joint to ensure a watertight connection.

Rising main

Once within the building the service pipe is usually referred to as the rising main.

Sacrificial anode

Protects cistern from corrosion. Consists of a lump of magnesium which dissolves when electrolytic action takes place. Thus the zinc lining will be protected because magnesium has a higher electric potential than zinc. (When electrolytic action takes place it is the metal with the higher potential that dissolves.)

Scale or fur

A deposit on the internal surfaces of boilers and pipes and on the external surfaces of immersion heaters in hard water areas.

S-trap

Trap with a vertical outlet.

Seating

The surface inside a tap or stop-cock on to which the tap washer presses to cut off the water.

Septic tank	A small underground chamber in which domestic sewage is liquefied by bacterial action. The first part of a small sewage plant.
Service pipe	The pipe taking the domestic water supply from the Water Authority's stop-cock into the house.
Single stack drainage	A system of above ground drainage in which all wastes discharge into one main soil and waste pipe.
Siphonic action	The flow of water from a cistern or trap as a result of atmospheric pressure.
Siphonic lavatory suite	A lavatory suite in which atmospheric pressure is used to empty and cleanse the pan. With these suites the flush pipe may be omitted and the pan and cistern 'close coupled' to form one unit.
Soil-pipe	The main above-ground drain-pipe into which lavatory wastes and, with a single stack system, other wastes discharge. This pipe is taken open-ended to above eaves level to serve as a ventilator for the drain.
Soldered capillary joint	A means of joining copper tubing which makes use of 'capillary attraction' – the tendency of any liquid (in this case molten solden) to flow to fill any narrow space between two surfaces. Often called

'Yorkshire joints' after one well-known brand.

Spacer washer	Another name for the 'top hat' washer used when fitting a pillar tap to a sink, bath or basin made of thin material. Also the washer that separates the grab ring and the 'O' ring seal in an acorn joint.
Spigot	The plain end of a pipe fitting into the socket of another.
Stabiliser	Device fitted to the arm of the ball float to prevent it from bouncing on every ripple.
Stand-pipe	An open-ended pipe fitted vertically into a cistern or tank to allow water to enter or leave. A stand-pipe forms the outlet of a Burlington flushing cistern and the hot water inlet of a galvanised steel hot water storage tank.
Stop-cock	A tap which enables the water supply to be cut off. In any plumbing emergency the first step is to turn the water off at the stop-cock.
Swaging tool	Special swaging tool which is inserted in the tube end and is turned full circle forcing a hard steel ball to make a groove round the inside of the pipe and an equivalent ridge or 'swage' round the outside.
Swivel tap connector (*cap and lining joint*)	Connects taps and ball valves to water supply pipes.
Trap	Device through which waste is

discharged designed to hold sufficient water to prevent drain smells re-entering the room.

Two pipe drainage The above ground drainage system in almost universal use up to the late 1950s. Soil appliances discharged directly into the drain or soil-pipe. Waste appliances discharged over a yard gully.

UDB water heater A squat hot water storage cylinder fitted with one or more electric immersion heaters and designed for under-draining-board installation to provide hot water supply by electricity.

Vermiculite chips Lightweight granules, widely used for insulating lofts.

Washer The small replaceable part of a tap which makes a seal on the tap seating and cuts off the water.

Water hammer Shock waves resulting from the sudden stopping of a flow of water. Consists of heavy knocking sounds following the closing of a tap or ball valve.

Water softening The removal, usually by ion exchange, of chemicals responsible for hard water.

Water waste preventer (WWP) A flushing cistern which gives a measured 9 litre (2 gal) flush.

Index

acorn fittings 162–3, 237
air locks 15, 36–8, 237
anti-siphon traps 112, 114
asbestos cement
  cisterns 11, 16
  rainwater systems 120, 122

back plate elbows 204–5
back plate tees 207
balanced flues 39
ball valves 13, 37, 48, 56–64, 210, 238, 242
baths 211, 212–18
bending pipes 168–9, 209
bib taps 45, 189, 193, 204–205, 238
bidets 15, 211, 227–32
blocked drains 126, 134–8
blocked waste pipes 116–17
boiler explosions 100–102
boiler scale 19–23, 43–4
burst pipes 10, 96–8

Calgon water softener 80

cap and lining joints *See* swivel tap connectors
cast iron rainwater systems 120
central heating 23–4, 87–9, 99–100
cesspools 143–7, 153–4
circulation 18–19, 27–31
cisterns
  cold water storage 8–16, 17, 19, 24–6, 32, 37, 82–5, 91, 92–4, 193, 239
  flushing 65–70, 71, 94–5
combined drains 132–4
communication pipe 5, 239
compression joints
  Type 'A' 97, 159–62, 171–172, 198–202, 203–207, 218
  Type 'B' 163–4, 171–2
condensation 9, 10, 75–6
copper tubing 5, 97, 158–70
corrosion 11, 23, 81–9
cup and cone joints 179–80, 240

cylinder collapse 103
cylinders, hot water
    direct 17–21, 26, 99
    indirect 21–4, 26, 99–100
    self-priming 23–4, 26, 89,
        240

dead legs 27, 240
dezincification 85–6, 162,
    240
direct cold water service 8–
    10, 24, 99, 241
direct hot water system 17–
    20, 241
distribution pipes 13, 14–15,
    91, 241
ditches 154, 156–7
drain-cocks 6, 52–3, 99, 241
drain leaks 139
drain smells 118, 129, 137–8
drain rods 135
drains
    above ground 104–23
    below ground 125–39

electric water heating 26–34,
    41–4
electrolytic corrosion 82–3,
    84–5, 87–9, 242
electro-thermal frost protec-
    tion 94–5
expanded polystyrene 76, 92

feed and expansion tanks
    21–3, 26, 87–9, 242
fibreglass 92
filter beds (drainage) 149–51
float valves *See* ball valves
flush panels 236
free outlet water heaters 21,
    41–3, 242

fresh air inlets 128–9, 138
frost precautions 5–6, 10,
    11, 13, 15–16, 25, 34,
    91–6, 98–100
frost stat 99, 243

garden water supply 45,
    203–7
gas water heating 38–44
gate valves 14–15, 28, 37,
    55–6, 92, 193, 243
glands and gland packing
    51–2, 55, 243
guard pipes 4, 243
gullies 107, 115–16, 118,
    134, 138

hard water 19, 43–4, 78–81,
    243–4
head-gear 46, 49, 244
hopper heads 107, 118, 244
hydraulic head 220, 244

immersion heaters 19–20,
    26–34, 99, 244
indirect cold water service
    8–10, 49, 244
indirect hot water systems
    21–6, 86, 99–100, 244–5
inspection chambers 125–
    132, 134–8, 155–6
instantaneous water heaters
    17, 39–41, 222
intercepting traps 128–9,
    135–8, 245

lagging 31, 32, 92–4, 103,
    245
lavatory pan (renewal) 232–
    235

lavatory suites
  close coupled 72
  high level 65–7, 236
  low level 67, 236
  siphonic 72–4, 248
  wash-down 70–2
lead in water supply 80–1
lead pipes 5, 80–1, 97–8, 172–80

magnetite 87, 245
Marley collar-boss 115, 245
metrication xiii, 159, 161, 173, 218, 228
Micromet crystals 20–1, 43–44, 86, 245
mixers
  bath and basin 47–8, 214, 224–5, 228, 245–6
  shower 48, 214, 221, 245–246
  sink 48, 190, 245–6
Multikwik connectors 75, 234–5, 236

noisy plumbing 9, 10, 12, 38, 57, 60–2, 67, 74, 89
non-return valve 207–209

off-peak water heaters 33–4
one-pipe circulation 30–31
one-pipe drainage 110–12
open outlet water heaters *See* free outlet water heaters
overflow or warning pipes 13, 94, 118, 246

packaged plumbing 24–6, 246

pillar taps 45–7, 190, 226–227, 246
pipe fixing 169–170
pitch fibre pipes 151, 155, 187–8
polypropylene 184
polythene 5, 185–7, 206–207
pop-up wastes 224, 230
pressure reducing valves 207
primary circuit 21–4, 86, 246
private sewer 133–4
PTFE thread sealing tape 172, 206, 246
public sewer 132–3
push-fit connectors 181

rainfall 119
rainwater drainage 119–23, 153
reversed circulation 28–30
ring seal joints 183–4
rising main 6, 91, 99, 246
rodding points 131

sacrificial anodes 84–5, 246
safety valves 102
saunas 211, 222–3
scale inhibitors *See Micromet* crystals
screwed iron pipes 172
septic tanks 143, 147–154, 248
service pipe 5–6, 91, 248
sewers 124, 132–4, 143
shower mixing valves *See* mixers, shower
shower pumps 220
showers 11, 15, 211, 218–22
silencer tubes 60, 62
single stack drainage 95, 111–16, 202, 225, 230, 248

sink waste disposal units
    196–8
sinks 189–98
siphonic action 66–9, 108–
    109, 248
snap ring joints 187–8
soakaways 124
soft water 79–81
soil fittings 106, 110
soil pipes 95, 106, 110, 230,
    248
solar heating 34–6
soldered capillary joints 97,
    165–8, 171–2, 248
solvent welding 181–3
spacer washers 162–3 *see
    also* top hat washers
stabilisers 60–1, 249
Staern Joints 177–9
stainless steel tube 171–2
stoneware drains 125–7
stop-cocks 4, 6–7, 53–5, 92,
    94, 100, 198–200, 203–4,
    206, 208–10, 249
subsoil irrigation 151–3
subsoil drainage 154–7
Supataps 45–7, 50
swivel tap connectors 162,
    194, 217, 225, 249

Taft joints 180–242
taps 45–52, 190–6, 203–7,
    214, 217–18, 221, 224–5,
    226–7, 228
thawing a frozen system 96
thermostats 20, 31, 221
Thru-flow valve 200–202
tools 1–3
top hat washers 191, 214
traps 105, 106, 110–12, 118,
    192–3, 216, 225, 249
two pipe drainage 106–10,
    225, 230, 249

U-can copperbend 169–70,
    209, 218, 225
UDB water heaters 32, 34,
    249
underground drains *See*
    drains, below ground
uPVC pipes 151, 155, 180–4
uPVC rain water systems
    122

valve seatings 45, 50–1, 59
vanity basins 223, 225
vent pipes 17, 21, 37–8, 106,
    108–10, 115

DO IT YOURSELF – A BASIC MANUAL

TONY WILKINS

Do you know the difference between a Pozidriv and a dowel screw? Do you know how to fix shelves, combat damp, replace a washer? This book will tell you how.

Do-It-Yourself is written for the absolute novice, by an author with over twenty-five years' experience in the DIY field, and is a beginner's guide to basic repairs in the home. It covers such jobs as replacing a plug or a fuse, and painting and wallpapering, often accompanied by explanatory diagrams. Advice is given too on buying materials and dealing with domestic emergencies.

Tony Wilkins is editor of 'Do It Yourself' magazine and has also written the very successful HOUSE REPAIRS in the Teach Yourself series.

TEACH YOURSELF BOOKS

CARPENTRY

CHARLES HAYWARD

This book is written for the man (or woman) who wants to take up woodwork as a hobby. It not only tells him exactly what he will want to know but also explains how to avoid the snags he will inevitably meet.

Through this book, the beginner will learn how to progress smoothly from the simpler to the more difficult carpentry jobs. With the use of working diagrams and illustrations, each stage is carefully explained to give the woodworker all the practical information he needs, along with a wealth of useful hints, plans and new ideas.

'Mr. Hayward gives you not only sound advice on tools and workshop practice, joints and their applications, indoor and outdoor woodwork, but also the designs and instructions for the making of a number of useful things.'

Ideal Home

TEACH YOURSELF BOOKS

HOUSE REPAIRS

TONY WILKINS

Buying a house is probably the biggest financial outlay you will ever make. However, if this asset is not to become a liability, a continuous programme of repair and maintenance is of great importance. Neglect soon leads to an increasing number of problems and general deterioration of the fabric.

The problems you meet will, of course, vary according to the type of property you have but, whatever they are, you can make considerable financial savings by dealing with them yourself.

This book will both help you to identify trouble spots and provide you with the knowledge to enable you to do a good job. A summary of the problems you might find, both inside and outside the house, is followed by a reference section of advice on how to tackle them. Topics covered include the roof, doors, windows, walls, damp, house surrounds, floors, ceilings, drainage.

Tony Wilkins is Editor of DIY magazine.

TEACH YOURSELF BOOKS